Safety of Biological Products Prepared from Mammalian Cell Culture

Developments in Biological Standardization

Vol. 93

This series Dev Biol Stand begins with Vol. 23 and is the continuation of both «Progress in Immunobiological Standardization, Vols 1-5» and «symposia series in Immunobiological Standardization, Vols 1-22».

KARGER
Basel · Freiburg · Paris · London · New York ·
New Delhi · Bangkok · Singapore · Tokyo · Sydney

Safety of Biological Products Prepared from Mammalian Cell Culture

Conference Centre of the Marcel-Mérieux Foundation
Les Pensières, Veyrier-du-Lac, Annecy, France
September 29-October 1, 1996

Volume Editors

Fred Brown
Plum Island Disease Center, Greenport NY, USA

Elwyn Griffiths
World Health Organization, Geneva, Switzerland

Florian Horaud
Institut Pasteur, Paris, France

John C. Petricciani
Palm Springs CA, USA

Proceedings of a Symposium organized and sponsored by
the Marcel-Mérieux Foundation and
the International Association of Biological Standardization

and co-sponsored by the World Health Organization

22 figures and 32 tables, 1998

KARGER

Basel · Freiburg · Paris · London · New York ·
New Delhi · Bangkok · Singapore · Tokyo · Sydney

••••••••••••••••••••••

Developments in Biological Standardization

Bibliographic Indices. This publication is listed in bibliographic services, including Current Contents® and Index Medicus.

Drug dosage. The authors and the publisher have exerted every effort to ensure that drug selection and dosage set forth in this text are in accord with current recommendations and practice at the time of publication. However, in view of ongoing research, changes in government regulations, and the constant flow of information relating to drug therapy and drug reactions, the reader is urged to check the package insert for each drug for any change in indications and dosage and for added warnings and precautions. This is particularly important when the recommended agent is a new and/or infrequently employed drug.

Printed in Switzerland by Médecine et Hygiène, Genève
ISBN 3-8055-6732-4

Acknowledgements

The International Association of Biological Standardization (IABS) wishes to record gratitude to the following companies, the institutes and organizations who have generously supported this Symposium.

AMGEN, INC, Thousand Oaks, CA, USA
ARES SERVICES SA, Geneva, Switzerland
CHIRON BIOCINE, Sienna, Italy
CENTEON MANAGEMENT LLC, King of Prussia, PA, USA
CENTOCOR, INC., Malvern, PA, USA
CONNAUGHT LABORATORIES LTD., North York, Ontario, Canada
GENENTECH, INC., South San Francisco, CA, USA
GENETIC THERAPY, INC., Gaithersburg, MD, USA
GENETICS INSTITUTE, Cambridge, MA, USA
GLAXOWELLCOME, Beckenham, Kent, U.K.
HOFFMANN-LA ROCHE, INC., Nutley, NJ, USA
MERCK & CO., INC., Rahway, NJ, USA
PASTEUR MERIEUX MSD, Lyon, France
PASTEUR MERIEUX S & V, Marcy l'Etoile, France
RHONE-POULENC RORER FOUNDATION, Antony, France
SANOFI RECHERCHES, Toulouse, France
SMITHKLINE BEECHAM, Philadelphia, PA, USA
SMITHKLINE BEECHAM PHARMACEUTICALS, Rixensart, Belgium
SWISS SERUM & VACCINE INSTITUTE, Berne, Switzerland
WYETH LEDERLE VACCINES AND PEDIATRICS, W. Henrietta, NY, USA

Contents

Brown F, Griffiths E, Horaud F, Petricciani JC (eds): Safety of Biological Products Prepared from Mammalian Cell Culture.
Dev Biol Stand. Basel, Karger, 1998, vol 93, p 1

Safety of Biological Products Prepared from Mammalian Cell Culture

Brown F, Griffiths E, Horaud F, Petricciani JC (eds): Safety of Biological Products Prepared from Mammalian Cell Culture.
Dev Biol Stand. Basel, Karger, 1998, vol 93, pp 3-4

Historical Background, Objectives and Overview

J.C. Petricciani

Genetics Institute, Inc., Cambridge MA, USA

Simply put, the major goal of this meeting is the resolution of several issues on the acceptability of a variety of cells used in the production of biologicals. Another, and-probably equally important goal is to begin to address new issues that are emerging related to-cells. Some of the old issues have been with us for more that three decades, while others have-been looking for a rational conclusion for only 10 years. And as Robin Weiss said last night, it is likely that new issues will arise for the foreseeable future.

It was my fortune, or misfortune, depending on one's perspective, to have started my biomedical career at the epicentre of the debate over human diploid cells almost 30 years ago. That seems like ancient history now, and in a sense it is. But, as I have pointed out many times, it is a useful history **only** to the extent that we learn from it and avoid repeating the same mistakes. This meeting, in fact, represents an attempt by WHO and the IABS to complete the resolution of several specific points on a global basis by taking into consideration not only the work that has been done in the past by National Control Authorities (NCAs), but also the more recent efforts by the International Conference on Harmonization (ICH) and-WHO itself. If we can come to agreement on taking definitive steps to resolve at least some of the remaining issues on cell substrates, based on the data available, we will have made an enormous step forward that should have a direct impact on the ICH activities in this area.

We are fortunate in having a significant background of information available for this meeting. Some of the information was developed during past meetings, but more recent data and summaries of specific areas were contributed by several of the participants who are with us today, as well as by others who could not be here. All this background material provides the basis for the specific proposals that are the subject of this meeting.

Several points merit special mention. First, one of the general themes of the revised WHO Requirements for Cell Substrates is to attempt to normalize or harmonize the requirements for primary cell cultures, diploid cell lines, and continuous cell lines. While some differences still remain, they have been minimized to the greatest possible extent. Of the more traditional requirements, karyology and tumorigenicity testing are two areas that have been revised extensively, based on experience and practical considerations. The more recent issue of residual cellular DNA has also been revised to lower its importance from that of being required as a release specification to that of a simple impurity with no special safety consideration. A similar case can be made for host cell proteins.

I will not take time to describe the background to those changes because they were given to you in the background information provided, and will be elaborated upon by the speakers today. Our aim for the background material and the presentations to be given by the speakers is to provide a sound overview of topical areas so that discussions can be focused on the critical issues about which there is legitimate uncertainty. Ultimately, we should leave this meeting having done our best to put to rest those issues that no longer need debate so that we can focus our collective attention on areas that are truly important. In that regard, Dr. Robin Weiss provided more than enough thoughtful insights on emerging potential issues to keep us busy for at least the next decade.

An astronomer described a recent meeting of the world's leading cosmologists as the most important gathering in years because their aim was to break down scholarly reserve and highlight the points of dispute and consensus. It was successful, again in the words of the astronomer who described the meeting, because people were willing to say the things they really believed to their colleagues at the meeting, rather than privately to their friends over a beer or glass of wine. It is our hope that that same atmosphere of openness and candour will characterize this meeting.

Along those lines, let me start off by saying that in all candour, if an intelligent being from another planet who had never considered cell substrate issues before looked at the requirements – both national and international – that now exist for various cell substrates, he (or she) would almost surely conclude that we are logically inconsistent because we are either demanding too much for diploid cells and continuous cell lines or too little for primary cells. If one looks back at the history of biological products, it is painfully clear that one issue stands out over and over again as the single most important one on which we must continue to be diligent: infectious agents. Whether it is polio vaccine, growth hormone, blood products, or very large scale biotech production, the recurrence of problems with infectious agents should be a potent reminder to all of us that from a public health point of view things such as cellular DNA and host cell protein are trivial considerations which have acted more as a distraction than anything else. Indeed, Phil Minor put his finger directly on the point recently when he wrote that DNA poses certain **psychological** regulatory problems. It is now time for the science of these matters to act as therapy for whatever psychological impediments there may have been in the past so that we can get on with the many new and exciting issues that are demanding our time and attention, and which do have real public health importance.

One final word: the agenda for tomorrow should be considered tentative in the sense that we can be somewhat flexible with respect to topics and time allotted. So, on behalf of the organizing committee, let me ask you to let Florian Horaud, Elwyn Griffiths, or me know if you feel that some adjustments should be made. Our objective in tomorrow's discussions is to provide an opportunity to develop a consensus wherever it is possible, and to identify areas where more work is needed to resolve issues.

Dr. J.C. Petricciani, P.O. Box 1925, Palm Springs, CA 92263-1925, USA.

Brown F, Griffiths E, Horaud F, Petricciani JC (eds): Safety of Biological Products Prepared from Mammalian Cell Culture.
Dev Biol Stand. Basel, Karger, 1998, vol 93, pp 5-13

Karyology and Tumorigenicity Testing Requirements: Past, Present and Future

J.C. Petricciani, F.N. Horaud

Genetics Institute, Inc., Cambridge MA, USA

The objectives of this paper are to: (i) review briefly the history of cell substrates as it relates to the introduction of karyology and tumorigenicity requirements; (ii) describe the current situation; and (iii) suggest what might be appropriate for the future.

The Past

The history of animal tissues and cells as substrates for the production of biological products dates back to the earliest efforts to test and manufacture biologicals [1-3]. The safety of biologicals produced in different cell substrates, however, did not emerge as an issue until the 1950s during the early days of inactivated polio virus vaccine production. The specific question at that time was whether it would be safe to use the human cancer cell line, HeLa, as a cell substrate [4]. The alternative was to continue using primary monkey kidney cell cultures. Concern was expressed about the potential for transmitting whatever it was in HeLa cells that endowed them with their capacity to generate tumours in the hamster cheek pouch, which was the only animal test system for tumorigenicity then available. The decision to reject HeLa cells and to endorse primary monkey kidney cell cultures was extremely important, not just for the issue at hand, but also because it established firm attitudes that would endure for decades regarding what types of cells would be acceptable, as has been noted previously by several authors [1, 3-5].

A different approach to the use of cells in vaccine production was made possible by studies in the late 1950s and early 1960s [6-8] which resulted in the serial propagation of human embryonic lung fibroblasts. Hayflick and Moorhead showed that such cells maintained a normal diploid chromosomal constitution until they entered senescence [8]. Such a cell system offered many potential advantages in that a standardized human diploid cell (HDC) line such as WI-38 could be used for many years with presumably predictable characteristics. But there were also objections raised regarding the use of WI-38 cells to manufacture vaccines for humans because of the theoretical possibility that they contained a human oncogenic agent for which no tests were available [9]. In that regard, the objections were the same as those already mentioned for neoplastic cells.

One of the most important initial meetings on HDCs was held in 1963 [10]. The major objective of that meeting was to emphasize the characteristics that differentiated HDCs from continuous cell lines (CCLs). Two major features were identified: (i) finite life for HDCs versus infinite life for CCLs; and (ii) a diploid chromosomal constitution for HDCs versus heteroploidy for CCLs. Those two characteristics were then considered together as descriptive of normal cells versus cancer cells. This was an extremely important point for the proponents of HDCs since the use of cancer cells for the production of biologicals had already been ruled out (4). While HDCs were clearly different from primary cells and CCLs, the question being debated was whether they more closely resembled normal primary cells or cancer cells.

One way of addressing that was to ensure that any HDC being considered as a substrate for biologicals was as close to normal as possible, and clearly different from cancer cells. Since it was well known that cancer cells display chromosomal abnormalities of number and structure, the cytogenetic criteria for accepting a HDC became very strict. Two examples in karyology serve to illustrate this. The occurrence of even a single definitely rearranged karyotype or abnormal chromosome, such as a dicentric, was enough to reject the HDC as a candidate cell substrate for a vaccine. Similarly, if tetraploid values of 8% or more among the metaphase cells were obtained, the HDC would have been rejected [11].

The first major international meeting on HDC substrate issues in the United States was held in 1967. Dogmatic positions, passionate arguments, as well as pleas for caution and a «go-slow» approach can be found in the proceedings of that conference [2]. Neither the papers presented nor the discussions really settled anything, and the debate over HDCs continued and intensified. At the heart of the controversy was the question of safety. Specifically, some individuals, such as Albert Sabin, expressed concerns that products derived from HDCs would carry a risk of transmitting viruses pathogenic for man, especially oncogenic viruses for which there were no tests. The answers to those safety questions were indirect and they did not persuade some opponents of HDCs because technology available at that time limited the assessment of HDCs to showing that they:(i) had a finite life; (ii) were predominantly diploid; (iii) contained no detectable viruses; and (iv) were not tumorigenic. In an effort to make a more convincing argument that HDCs were safe, the cells were tested repeatedly for senescence, cytogenetic stability, viral contaminants, and tumorigenicity. So many of those tests were done in various laboratories around the world with consistent results that those characteristics of HDCs became the new standard of acceptability.

Karyology

In 1968, another meeting on HDCs was held in Zagreb to discuss the oncogenicity of virus vaccines [12]. At that meeting, criteria were developed for the acceptability of a diploid cell population as a substrate for virus vaccine production, and the rationale for karyology was stated as follows [13]: «Karyological characterization of any diploid cell strain proposed for use in the production of vaccines for human use is necessary in order to detect gross aberrations in chromosome number or morphology which might indicate:

a) Abnormalities intrinsic in the original foetal material.

b) Possibilities of mislabelling and/or contamination involving established cell lines.

c) The presence of micro-organisms such as viruses, mycoplasmas, etc.

d) The inadvertent exposure of the cell strain to chemical or physical agents which might similarly induce persisting chromosomal changes.»

As noted above, particular emphasis was placed on the normal karyology of HDCs as an indication of their equivalence with normal cells and thereby their safety. A large pool of chromosomal data were accumulated, and standards of acceptability were established [14]. Recommendations were also developed for safety tests during the production of vaccines in WI-38 cells, and they included karyology monitoring of every lot of cells. Such monitoring in turn led to the development of an even larger body of cytogenetic data on WI-38 cells which tended to confirm the overall view that the cells displayed only low levels of chromosomal abnormalities until senescence.

Beyond HDCs

The next major step in the evolution of cell substrate issues came in the mid-1970s. At that time there was a great deal of interest in the use of interferon (IFN) in various clinical studies. Rather large amounts of IFN could be produced using the Namalwa human lymphoblastoid cell line, and those cells were proposed as a source of IFN for human clinical studies by the Wellcome Research Laboratories [1]. IFN from malignant human lymphoid cells presented serious questions of safety and provided the first direct challenge to the rather firmly established concept that acceptable cell lines had to possess HDC characteristics of a finite life, a normal cytogenetic profile, and not produce tumours. A conference was held at Lake Placid in 1978 to discuss the issues posed by Namalwa IFN and to open a reconsideration of the entire area of cell substrates [15]. The conclusion of that meeting was a cautious «yes» to the continuing exploration of lymphoblastoid IFN as an experimental biological product. Nevertheless, there was a certain amount of ambivalence among the participants in the same way that there was when HDCs were first being considered.

Interest then began to be expressed in the use of other CCLs for other products such as inactivated polio vaccine using Vero cells. Subsequently, biotechnology led many groups to pursue CCLs such as CHO as expression systems for new products and hybridomas for monoclonal antibodies. However, there was still a great deal of concern relating to the safety of products for human use derived from various CCLs. Another major international conference on cell substrate issues was held in 1984 during which it was suggested that, rather than concentrate on cell characteristics such as tumorigenicity and karyology, attention should be focussed on risks that might be associated with three potential contaminants from the cell substrate: (i) viruses; (ii) DNA; and (iii) transforming proteins [16]. There was general agreement with that approach, and the conclusion of the meeting was that CCLs should, in principle, be acceptable as substrates for biologics production, but the specifics of the manufacturing process should be the determinants in deciding on the acceptability of any given product derived from a CCL. Those same principles were reinforced by the WHO Study Group on biologicals in 1986 [17].

During the period 1960 to 1980, developments in the veterinary vaccine field were taking place rather independently of those in human biologicals, although the safety of veterinary products was also of major concern. Specifically, the use of BHK-21 cells for producing foot-and-mouth disease vaccine and rabies vaccine was proposed because of the significant advantages that cell system could provide [18]. After a period of controversy, and the generation of data in support of the safety of vaccines derived from BHK-21 cells, there was rapid movement towards the use of that cell system. At the present time, hundreds of millions of doses of vaccines are produced in BHK-21 cells and used each year [19].

The Present Situation

Karyology

Karyology requirements, both national and international, apply only to biological products derived from diploid cells. This is the result of the fact that when they were originally introduced, the use of HDCs to produce biologicals was new, and there was concern that some cell cultures might spontaneously transform because of their inherent genetic background or because of contamination with a transforming virus or an already transformed (heteroploid) cell line. Chromosomal monitoring was reasonable as a check on such an event since one would expect to see cytogenetic abnormalities accompanying a change to the neoplastic state. The large experience of many laboratories now makes it clear that normal human cells do not transform spontaneously in vitro to neoplastic cells. Indeed, it has been extraordinarily difficult to transform human cells with chemical carcinogens. Another reason which has been suggested for doing chromosomal analysis on each batch of cells is that it is useful in detecting cells which have been exposed to irradiation, certain chemicals, viruses, and mycoplasma infections [13]. There is ample evidence in the literature that at least some mycoplasmas can induce chromosomal abnormalities, as can a number of other microbial agents. Similarly, it is now well recognized that irradiation and some chemicals are potent clastogens. The pertinent question, however, is not whether various agents can induce chromosomal abnormalities; but whether or not karyology is a reasonable approach for the detection of extraneous biological agents in cells, or for the detection of cells that have been exposed to physical or chemical agents.

There is also an issue of regulatory consistency. The use of cells which are chromosomally normal during the first part of their life, but which develop abnormalities as they age, and which may even become CCLs, has considerable precedence even though it may not be recognized as such. Primary cultures of chicken embryo, duck embryo, rabbit kidney, dog kidney, and monkey kidney cells have been used to produce hundreds of millions of doses of live virus vaccines without regard to the eventual fate of those primary cultures in terms of showing that they maintained diploid characteristics or that they entered into a senescent phase and eventually died. Our experience, as well as that of many other laboratories, clearly shows that such primary cultures may not be diploid in the strict sense of the term, and that some may indeed go on to develop chromosomal abnormalities [20]. However, we have no indication that the use of such cell systems has imposed risks or hazards on the populations receiving vaccines produced in them. On the contrary, the general safety of vaccines produced in primary cells is well established. The only safety issue that has emerged over the years relates not to the cells themselves, but to viral contaminants such as SV40. In large part, potential viral contaminants have been controlled by careful screening and in many cases by the use of colony-bred animals. Since primary cell cultures, by definition, are not associated with a cell bank, if testing for karyology and tumorigenicity were to be required for their use as cell substrates, they would have to be done on each production lot of cells. Such a requirement is not only unjustified, based on past experience, but impracticable since it would greatly increase the cost and/or restrict market supply with a net negative effect on public health.

Experience has shown that, in some cases, the later passages of primary cultures will become CCLs with chromosomal abnormalities. Vero, BSC-1 and LLCMk-2 cells are just three examples derived from primary monkey kidney cultures, which are still being used in the production of poliovirus vaccine in the United States, Europe, and Asia. Thus, for many years we have been using cells in vaccine production some of which have had the clear potential for becoming heteroploid with an infinite life span.

Even in the case of HDCs, strict diploidy is not maintained during their entire life history because in the later stages of their in vitro life, chromosomal abnormalities show a significant increase in frequency. However, those cytogenetic changes are not associated with any of the phenotypic characteristics of cancer cells (e.g. infinite life, tumorigenicity). Spontaneous mutations also occur with some frequency in cultured HDC lines, as well as in other cell lines. While chromosomal analysis is the means by which these genetic changes have been detected in HDCs, it is important to question the relevance of such analyses in the control of biological products.

Tumorigenicity

Tests of cells used in vaccine production for freedom from tumorigenic potential are currently required in the United States only to establish the original acceptability of HDCs and CCLs. Tests on each batch of human cells for production are required by British Regulations. No National Control Authority requires such tests for any of the primary cell cultures. Again, a double standard exists, the origin of which is found in the thinking that was prevalent in the 1960s regarding the safety of primary cultures and the potential risks of HDCs.

Tumorigenicity tests can be justified in the characterization of a new cell line which might be proposed for use in biologics production. However, routine tumorigenicity tests on each production batch of cells can be criticized on two counts. First, since any new cell line will have been characterized regarding not only its tumorigenic potential but the presence of latent agents and chromosomal constitution, to repeat tumorigenicity tests on every batch of such cells is a redundant and futile exercise. The second argument against the routine use of tumorigenicity tests relates to the sensitivity of the available tests and the basic question that is being asked. If the question is whether or not the diploid cell culture has been contaminated and contains a mixture of «normal» and neoplastic cells, then tumorigenicity testing is not the method of choice to make that determination. For example, most current tests are able to detect 10^6 tumour cells from 65-100% of the time when the endpoint is a gross tumour nodule at the site of inoculation after two to four weeks. Tumour cell lines vary in their tumorigenic potential as does the sensitivity of various animal test systems. A one to two log increase in sensitivity can be obtained if histopathological evidence of tumour cells is taken as the endpoint. Even then, however, a tumour cell contaminant of low tumorigenic potential would be difficult to detect unless it represented a significant percentage of the inoculum. Such gross contamination of an HDC culture by tumour cells can be detected much more readily by a chromosomal screening analysis than by tumorigenicity testing, if, as is almost always the case, the tumour cells deviated significantly from the diploid state. Even easier and much more sensitive approaches to identifying a tumour-cell contaminant are available based on enzymatic, immunological, and molecular biological methods, and they are clearly preferable to the relatively insensitive in vivo tumorigenicity test systems.

Future Directions

Taking into account the historical background from which the current requirements for karyology and tumorigenicity evolved, as well as both current use of various cell types and contemporary data, a reassessment of the appropriate role of those tests is reasonable and warranted.

Karyology

In view of the net experience with human cells in general, and WI-38 and MRC-5 cells in particular, it is time to reassess the need for the extensive chromosomal monitoring on **each production lot** of cells. When the original rationales for the karyology requirement are examined in the light of current knowledge and practice, it is clear that karyology is an outdated test that does not meet real needs. Specifically, karyology is not the method of choice for the detection of any known microbial agent since sophisticated and specific tests are available for their isolation and identification. Moreover, to justify the use of karyology to detect cells exposed to physical or chemical clastogens, there should be some recognizable opportunity for such events to occur. With reliable quality control procedures and good manufacturing practices in place, the procedures in the cell culture and media production areas essentially preclude such events, thus obviating the need for routine karyologic evaluation simply to detect such occurrences. Here again, other more contemporary approaches to quality control are available and preferable.

Perhaps the single most important contribution of the diploid cell system is that it established the concept of a well-characterized substrate prior to its use in biologics production. It is long overdue for us to take better advantage of this feature of diploid cells by removing the requirement for lot-by-lot chromosomal monitoring, and instead focus attention on the identity and purity of the cells to be sure they are in fact the well-characterized acceptable substrate they are supposed to be. There is a variety of enzymological, immunological, and molecular biological tests to assure the identity and purity of well-characterized cell substrates, all of which are superior to karyology as a tool for making those assessments.

Tumorigenicity

Tumorigenicity tests may be useful and appropriate in the characterization of a new cell line; but they are of questionable value when applied routinely because of their lack of sensitivity, and more importantly because of the expense and the waste of animals and time in repeated testing of a cell line whose properties are already known. Tumorigenicity tests should be deleted as a routine quality control test for HDCs.

Testing based on product type and risk assessment

The biological products with which we have had the greatest experience are the virus vaccines produced in cell culture systems. There are, however, three other types of biological products which must be considered in any discussion of the factors affecting the acceptability of cells and the quality control tests that might be appropriate. First are the non-replicating biological products which are synthesized by cells, and exemplified by interferon. The number of such protein by-products of cells is potentially very great. The second new type of cell culture product, which is in an earlier stage of development, is the cell culture itself. Specific functional cells, such as insulin-producing islet cells grown in vitro, are now being used experimentally to treat a variety of diseases. Cell transplants are now emerging as potentially important biological products. The third product category includes gene therapy vectors which resemble live viral vaccines produced in CCLs. In some cases, vector-containing cells are implanted into the patient. An important question raised by the three new classes of products is whether or not the same criteria of acceptability that now apply, or may apply in the future, to vaccine cell substrates also should be used for other classes of products. An examination of these three classes of products suggests that there is a clear difference in risk among them based solely on the consideration of the cell substrate involved. For example, the risk of using interferon produced from Namalwa cells is much less than that of transplanting Namalwa cells themselves into an immunosuppressed human. The reasons for the differences in risk relate directly to the nature of the final product plus the degree to which it can be purified and tested, in addition to the characteristics of the cells from which it was derived. The theoretical risks identified above become irrelevant to a product such as interferon if it can be shown that the purification process efficiently and consistently reduces contaminants to acceptable levels. This scheme proposes that any new cell line (HDC or CCL) should be subjected to the same general characterization studies regardless of the product class in order to establish a general overview of the cell system. Differences in the extent of testing at the stage of routine monitoring should correspond to the degree of risk (theoretical and real) associated with use of the product.

Summary and Conclusions

At the present time, karyology and tumorigenicity are applied to primary, diploid, and continuous cell systems in an uneven and inappropriate manner, largely for historical reasons. It is a significant anomaly that such rigorous requirements are applied only to diploid cells when, of all three cell types, they represent the category with the least potential problems. During the 1992 presidential campaign in the U.S., a very direct and telling slogan was used to the advantage of Mr. Clinton: «It's the economy, stupid». That message was meant to have the effect of focusing on the **real** issue that was of concern to the electorate.The other topics were of secondary significance and tended to act as distractions from the central issue. In a very similar sense, one could say: «It's the product, ..., not the cells». In other words, we should be focussing our time and attention on the characteristics of the biological product manufactured in a given cell system that has been well characterized rather than continue to belabour the issue of cell substrates. To a large degree, this has already been initiated with the recommendations of various groups already mentioned [16, 17]. However, current diploid cell quality control regulations stand out as a peculiar throwback to an earlier era. HDCs should be treated on a par with primary and continuous cells.

There are three basic questions related to the **routine use** karyology that need to be addressed: (i) is the original rationale for requiring cytogenetic analysis of a diploid cell substrate still valid; (ii) is there a new rationale that would warrant its continued use; and (iii) if there is a continuing need for karyology, is it unique to diploid cell cultures or does it need to be extended to all types of cell systems?

The original rationale was that karyology provided evidence of the normal character of WI-38 cells and therefore supported its acceptability as a cell substrate for vaccine production. Karyology essentially has remained as a legacy of the intense debate that led to the acceptance of WI-38 cells. There is no new information over the past 30 years to suggest that there is a new rationale for instituting chromosomal analysis of cell substrates. If there were, however, it is difficult to imagine why it should not be applied to all types of cell substrates (primary, diploid, and continuous). Taking all the above into consideration, there would seem to be no rationale for continuing to single out diploid cell cultures as the only cell type for which karyology is required on a **routine** basis. As stated above, the initial characterization of a **new** diploid cell line should include karyology.

Like karyology, tumorigenicity testing was incorporated into the assessment of diploid cells in an attempt to persuade regulatory authorities that the cells were normal and acceptable. Again, after 30 years of testing, there has never been an instance of normal diploid fibroblasts generating a tumour in any in vitro or in vivo assays. The futility of continuing to do these tests is obvious. A description of the tumorigenic potential of a cell substrate should be an element in the characterization of a new cell line; but it has little if any value as a routine test.

If history teaches us anything at all about risk, it is that we need to focus serious attention on contaminants rather than be diverted to remote theoretical issues that may be interesting to discuss and argue about, but which pale in the face of the potential impact on public health of viral and viral-like contaminants of biological products. One has only to recall the transmission of SV-40 from primary monkey kidney cells that were used to produce polio vaccine, or more recently the transmission of Creutzfeld-Jacob disease to recipients of human growth hormone derived from human pituitaries and of HIV to recipients of blood and blood derivatives. Indeed, the message that rings out loud and clear from past experience is that human tissue provides a very high risk source for transmitting infectious agents to recipients of biological products. By focusing on the biological product and using state-of-the-art technology to evaluate potential safety issues, it is possible to provide a high level of assurance that those products are safe, especially with regard to infectious agents.

REFERENCES

1 Beale AJ: Choice of cell substrate for biological products. Advances in Experimental Biology and Medicine 1979;118:83-97.

2 National Cancer Institute Monograph (USA) 1968;29.

3 Petricciani JC: Cells, science and health, in Hayflick L, Hennessen W (eds): Continuous Cell Lines as Substrates for Biologicals. Dev Biol Stand. Basel, Karger, 1989, vol 70, pp 3-10.

4 Hilleman MR: Cell line saga – an argument in favor of production of biologics in cancer cells. Adv exp Med Biol 1979;118:47-58.

5 Hull RN: Immunization of experimental animals with vaccines produced in known oncogenic cells. National Cancer Institute Monograph (USA) 1968;29:503-509.

6 Swimm HE, Parker RF: Culture characteristics of human fibroblasts propagated serially. Am J Hyg 1957;66:235-243.

7 Tjio JG, Puck TT: Genetics of somatic mammalian cells II. Chromosomal constitution of cells in tissue culture. J Exp Med 1958;108:259-268.

8 Hayflick L, Moorhead PS: The serial cultivation of human diploid cell strains. Exp Cell Res 1961;25:585-621.

9 Sabin A: Discussion. Session V of the International Conference on Rubella Immunization. Amer J Dis Child 1969;118:378-381.

10 Proceedings of the Symposium on Human Diploid Cell Strains. Opatija, Yugoslavia, 1963, vol 8.

11 Proceedings of the Symposium on Human Diploid Cell Strains. Opatija, Yugoslavia, 1963, vol 8, p 711.

12 Proceedings of the Symposium on Oncogenicity of Virus Vaccines. Zagreb, Yugoslavia, 1968.

13 Proceedings of the Symposium on Oncogenicity of Virus Vaccines. Zagreb, Yugoslavia, 1968, p 174.

14 Moorhead PS, Nichols WW, Perkins FT, Hayflick L: Standards of karyology for human diploid cells. J Biol Stand 1974;2:95-101.

15 Petricciani JC, Hopps HE, Chapple (eds): Cell Substrates. New York, Plenum, 1979.

16 Hopps HE, Petricciani JC (eds): Abnormal Cells, New Products, and Risks. Tissue Culture Association, Gaithersburg, 1985.

17 Report of the meeting of the study group on biologicals. WHO Technical Report Series, 1987, N° 747.

18 Rodericks JV: Annals of the New York Academy of Science 1981;63:29.

19 Regan PJ, Petricciani JC: The approach used to establish the safety of veterinary vaccines produced in the BHK-21 cell line, in Petricciani JC, Hennessen W (eds): Cells, Products, Safety. Dev Biol Stand. Basel, Karger, 1987, vol 68, pp 19-25.

20 Wallace RE, Vassington PJ, Petricciani JC, Hopps HE, Lorenz DE, Kadanka Z: Development and characterization of cells lines from subhuman primates. In Vitro 1973;8:333-341.

Dr. J.C. Petricciani, P.O. Box 1925, Palm Springs, CA 92263-1925, USA.

Brown F, Griffiths E, Horaud F, Petricciani JC (eds): Safety of Biological Products Prepared from Mammalian Cell Culture.
Dev Biol Stand. Basel, Karger, 1998, vol 93, pp 15-19

Cell Line Banking and Authentication

R.J. Hay

Cell Biology, American Type Culture Collection, Rockville, MD, USA

Seed Stock Concept

Definitions of public repository seed stocks may vary from those used for specific applications such as the production of vaccines or other biologicals. A scheme illustrating the steps involved in developing the seed stocks is presented in Figure 1.

Generally, starter cultures or ampoules are obtained from the originator and progeny are propagated according to instructions to yield the first «token» freeze. Cultures derived from such token material are then tested for bacterial, fungal, and mycoplasmal contamination. The species of each cell line is verified. These quality control steps are the minimum ones that must be performed before eventual release of the line. If these tests confirm that further efforts are warranted, the material is expanded to produce the seed and distribution stocks. Under ideal conditions, additional major quality control and characterization efforts are applied to cell populations from seed stock ampoules. Test results refer to **specific numbered stocks**. The distribution stock consists of ampoules that are distributed on request to investigators. The reference seed stock, on the other hand, is retained to generate further distribution stocks as the initial distribution stock becomes depleted. The degree of characterization applied to master cell banks or master working cell banks in production facilities is generally more rigorous, although the seed stock, like the master cell bank, is used as a reservoir to replenish depleted distribution lots over the years. By adherence to this principle, problems associated with genetic instability, cell line selection, senescence, or transformation can be avoided.

Authentication Steps

Bacteria and fungi – In testing for adventitious agents in cell cultures, bacterial and fungal contaminations represent a major concern. In most, but unfortunately not all instances, they are overt and easily observed. The inclusion of a series of culture tests permits detection of most common species that will thrive in cell-culture systems. Specifically, one can use a regimen of recipient cultures such as blood agar, thioglycollate, trypticase soy, brain heart infusion, Sabouraud and YM broths plus nutrient broth with 2% yeast extract. These are incubated at 37°C and 26°C, both aerobic and anaerobic

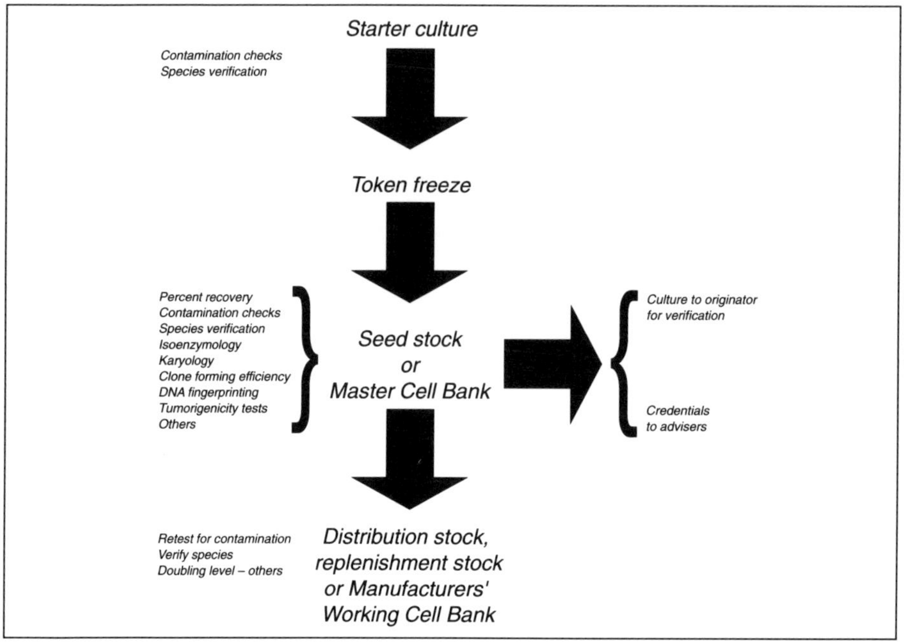

Fig. 1: A quality control scheme suggested for the addition (accession) of a cell line to a collection. Representative tests for token, seed, and distribution stocks are indicated (left), as are steps for review by originating investigators and advisers (right).
Terminology for the cell stocks may vary with the institution and specific bank utility. In production facilities the «seed stock» is generally referred to as the Master Cell Bank (MCB) and the «distribution or replenishment stock» may be defined as The Manufacturers' Working Cell Bank (MWCB).
Additional characterizations such as general or specific screens for viruses, functional tests, and definitions of particular inclusions or ultrastructure may be applied when critical, or with special support. Immunological tests are required, especially to identify specific protein inclusions or other products such as immunoglobulins released by hybridomas.

conditions are included for the blood agar and observation periods of 14-21 days are used. More methodological detail are provide in 21 CFR 610.12 and [1].

Mycoplasma – Contamination by mycoplasma is much more serious due to the insidious nature of such infection. The presence of some mycoplasma species may produce cytopathic effects but others may metabolize and proliferate extensively without inducing noticeable morphological changes in the host-cell culture. Still, the unrecognized introduction of such mycoplasmal infection interferes with studies on metabolism, receptors, virus-host interactions, cell division, and so forth, virtually negating research findings entirely. This problem in the cell-culture field has been emphasized repeatedly.

Table 1 summarises results from five different testing laboratories. While the periods of time, countries involved, and number of samples included vary considerably, it is clear that a significant problem exists internationally. Frequencies of infected cultures occured over a range of 5 to 64 percent.

Table 1: Mycoplasma infection of cell lines.

Reference test laboratory	Percent positive	Total
McGarrity, 1982 [2]	4.7	16.197
Del Guidice and Gardella, 1984 [3]	11.4	34.697
Gignac et al, 1991 [4]	64	39
Takeuchi et al, 1993 [5]	21	2.332
Lundin and Lincoln, 1994 [6]	10.2	1.000

Results of a testing service performed in our laboratory are summarized in Table 2. Note that the frequency of positive cultures varied from about 12 to 21 percent. We do not believe the trend is necessarily declining as the deviation from year to year is substantial.

Established methods for detecting mycoplasma in cell cultures are numerous including direct cultivation, indirect demonstration by fluorochrome staining (e.g. Hoechst or DAPI (see CFR 610.30) DNA hybridization and most recently the sensitive Polymerase Chain Reaction (PCR) [7, 8]. The latter, in use at the ATCC, requires primers based on the DNA sequences in 16S and 23S mycoplasmal rRNA. These amplify DNA from all the common mycoplasma found in cell cultures to levels easily detected after gel electrophoresis and ethidium bromide staining. This technique is at least as sensitive as the Hoechst stain, able to detect all the common species of mycoplasma, including *A. laidlawii* known to infect cell cultures. Furthermore, it does not suffer from the disadvantage in interpretation difficulties associated with some with Hoechst or DAPI-stained preparations [8].

Detection of viruses – These are certainly the most problematical of test series applied for authentication. A few of the more common techniques include examination for cytopathology, inoculation on to the chorioallantoic membrane (CAM) of embryonated chiken eggs, co-cultivation and more recently, DNA hybridization and PCR analyses. This subject will be covered by subsequent speakers.

Table 2: Frequency of mycoplasma infection in cultures tested* as a service at ATCC.

Year	Total number	% positive
1989	834	21
1990	958	13
1991	818	13
1992	821	18
1993	1031	16
1994	900	12

* Combination of direct culture and Hoechst staining used.

Cellular cross contamination – The problem of cell line cross contamination is also severe. Data developed by groups offering identification services indicate that as many as 17 to 35% of cultures submitted are not of the species or specific identities expected [9, 10]. Representative data are summarized in Table 3.

Techniques in wide use for cell-cell cross contamination detection include immunotyping, iso-enzymology, cytogenetic analysis after Giemsa banding and, more recently, DNA fingerprinting [1, 11-14]. PCR amplification of hypervariable regions for DNA profiling has also been informative in cell line identification through use of primers such as ApoB, D1S80 and D17S5 [15]. Fragments of DNA of differing sizes, thus visualized, provide unique profiles for each cell line. The new method is advantageous in that it requires very little material and is comparatively rapid and informative.

Two disadvantages of the molecular techniques for cell identification are: (i) their lack of ability to detect very low levels (i.e. less than 1%) of contamination and (ii) the absence of extensive background data. By comparison, cytogenetic evaluations are already available for many cell lines in common use. Furthermore, by examining many metaphase spreads, the expert cytogeneticist may recognize subpopulations within the parent line plus contamination by other lines at levels of 1% or less in some circumstances. Low levels of contamination may also be detected readily by the application of histo-enzymological stains specific for the lines in question [16].

Table 3: Summary of cell line cross-contamination[a].

Reported cell species	Cultures received	Interspecies	Intraspecies	Percentage of total
Human	160	40 (25%)	18 (11%)	36
Others	115	38	–	33
Total	275	78	18	35

[a] Adapted from B. Hukku et al (1984).

References

1 Hay RJ: Caputo J, Macy ML: ATCC Quality Control Methods for Cell Lines, 2nd ed., Rockville, MD, ATCC, 1992.

2 McGarrity GJ: Detection of mycoplasmal infection of cell cultures. Adv Cell Culture 1982;2:9-131.

3 Del Giudice RA, Gardella RS: Mycoplasma infection of cell culture: Effects incidence and detection. In Vitro Monograph 5, Uses and Standardization of Vertebrate Cell Cultures. Gaithersburg, MD, Tissue Culture Association, 1984, pp 104-115.

4 Gignac SM, Brauer S, Hane B, Quentmeier H, Drexler HG: Elimination of mycoplasma from infected leukemia cell lines. Leukemia 1991;5:43-53.

5 Takeuchi M, Yoshida T, Satoh M, Kuno H, Ohno T: Survey of mycoplasma contamination in animal cell lines collected by three cell banks in Japan. Bull JFCC 1993;9:13-18.

6 Lundin DJ, Lincoln CK: Mycoplasma testing of cell cultures by a combination of direct culture and DNA-fluorochrome staining. In vitro 1994;30A:11.

7 Harasawa R, Uemori T, Asada K, Kato I: Sensitive detection of mycoplasmas in cell cultures by using two-step polymerase chain reaction, in Kahane I, Adoni A (eds): Rapid Diagnosis of Mycoplasmas. New York, Plenum Press, 1993.

8 Hu M, Buck C: Application of polymerase chain reaction technique for detection of mycoplasma contamination. J Tissue Cult Meth 1993;15:155-160.

9 Nelson-Rees WA, Flandermeyer RR, Hawthorne PK: Banded marker chromosomes as indicators of intraspecies cellular contamination. Science 1974;184:1093-1096.

10 Hukku B, Halton DM, Mally M, Peterson WD, Jr: Cell characterization by use of multiple genetic markers, in Acton RT, Lynn JD (eds): Eukaryotic Cell Cultures. New York, Plenum Publishing Co., 1984, pp 13-31.

11 Gilbert DA, Reid YA, Gail MH, Pee D, White C, Hay RJ, O'Brien SJ: Application of DNA fingerprints for cell-line individualization. American Journal of Human Genetics 1990;47:499-514.

12 Honma M, Kataoka E, Ohnishi K, Ohno T, Takeuchi M, Nomura N, Mizusawa H: A new DNA profiling system for cell line identification for use in cell banks in Japan. In Vitro Cell and Dev Biol 1992;28A:24-28.

13 Hane B, Trummler M, Jager K, Schleithoff L, Janssen JWG, Drexler HG: Differences in DNA fingerprints of continuous leukemia-lymphoma cell lines from different sources. Leukemia 1992; 6:1129-1133.

14 Stacey GN, Bolton BJ, Doyle A: DNA fingerprinting transforms the art cell authentication. Nature 1992;357:261-262.

15 Reid YA, Luo X: The use of PCR-amplified hypervariable regions for the identification and characterization of human cell lines. In Vitro 1993;29A:120.

16 Markovic O, Hay RJ, Steenbergen K: Rapid screening and detection of cellular cross-contamination in cell cultures: A new application of cytochemistry. Histochem and Cytochem 1996; 44(1):75-76.

Dr. R.J. Hay, Cell Biology, American Type Culture Collection, 12301 Parklawn Drive, Rockville, MD 20852, USA.

Brown F, Griffiths E, Horaud F, Petricciani JC (eds): Safety of Biological Products Prepared from Mammalian Cell Culture.
Dev Biol Stand. Basel, Karger, 1998, vol 93, pp 21-29

Raw Materials as a Source of Contamination in Large-Scale Cell Culture

R.L. Garnick

Genentech, Inc., South San Francisco, CA, USA

Key words: Cell culture, investigation, minute virus of mice, MVM, PCR, raw materials, viral contamination.

Abstract: Large-scale cell culture operations for biotechnology products use millions of litres of complex media and gases as well as huge quantities of organic and inorganic raw materials. These raw materials must always be assumed to contain contamination by adventitious agents such as Minute Virus of Mice (MVM). Genentech has had experience in dealing with two such contaminations. Although the source of these contaminations was not positively identified, there was strong evidence to suggest that the virus entered through a raw material used in cell culture. Analytical methods that can be used to detect the presence of viruses can be used as an early warning system and as part of a strategy to devise barriers against such contamination. Methods such as a polymerase chain reaction (PCR) assay and a cell culture-based infectivity assay have been found to be efficacious in providing early detection of MVM infection. In any contamination, timely interactions with regulatory authorities are vital in minimizing delays in product manufacture.

Introduction

Large-scale cell culture contamination of recombinant DNA (rDNA) products by adventitious agents such as viruses and the role of raw materials as a source of such contamination must be understood in the proper perspective when considering the impact of such contaminations. The earliest rDNA products were produced in *Escherichia coli* bacteria and included human growth hormones and interferons as well as insulin. There was clear recognition during the development of the fermentation systems for these products of the need to avoid contamination by bacteria, bacteriophage, and yeast/fungi. Significant advances were made in the design of fermenters and in the development of manufacturing practices and procedures to minimize the chances of such contaminations. Nevertheless, contaminations occasionally did occur and often could be traced to either equipment failures and/or personnel. The development of contained fermenter systems was effective in limiting the spread of these contaminations. The contaminations themselves were generally very apparent and easy to detect by standard microbiological techniques.

As knowledge of biotechnology expanded, so did the industry's ability to produce such large glycosylated proteins as tissue plasminogen activator (t-PA), DNase, and Factor VIII. These molecules were produced in cell lines (CHO-K1, BHK-21) in cell culture facilities capable of handling up to 10,000 litres of material. Since the early 1980s, these facilities have used countless millions of litres of raw materials and media. This one factor alone is significant in the prediction of contamination by adventitious agents. It was apparent from the beginning of this technology that viral contamination of cell cultures from these sources was a real threat. Therefore, extensive measures were introduced in an attempt to avert potential contamination. An exhaustive characterization of the master cell banks (MCB) or working cell banks (WCB) for endogenous viruses (retroviruses) as well as for other adventitious agents was performed together with routine testing of every lot of cell culture for mycoplasmas, viruses, and bacteria/fungi. These programmes resulted in a very favourable track record with regard to contaminations, which were relatively few and usually bacterial or fungal in nature. In addition, no retroviral infectivity was observed although reverse transcriptase activity was detected. Only sporadic and often unconfirmed reports of adventitious viral contaminations were noted, although MVM was previously observed by others [1].

The difficulties inherent in detecting low levels of viral contaminants and thus the potential to unknowingly contaminate a whole manufacturing facility lie at the centre of the concerns about viral contamination (Fig. 1).

Vulnerability Considerations

The worldwide use of rodent-derived cell lines such as CHO-K1 and BHK-21 for biotechnology products emphasizes the scope of the potential for viral contamination of the cell lines for these products. Literally millions of litres of complex media, air and other gases, together with tons of organic and inorganic raw materials have been used in the manufacture of these products. Experience has indicated that the quality control system for any of these products must be built on the assumption that any raw material may be contaminated with a virus. The potential sources of viral contamination in large-scale cell cultures are presented in Figure 2, demonstrating the extent of the problem in identifying the source of a viral contamination from the myriad possibilities.

Viral Contamination

- Historical and continuing problem in cell cultures
- Viruses are not always easily detected without amplification
- Viruses are small particles that are not easily removed by filtration
- Viruses are difficult to inactivate and potentially pathogenic
- Regulatory authorities must be notified in the event of a contamination

Fig. 1: Concerns regarding viral contamination of large-scale cell cultures.

Experience With MVM

Genentech has experienced two large-scale cell culture contaminations by MVM, whose characteristics are presented in Figure 3. These contaminations resulted in considerable expense and delayed product manufacture. The two contaminations were completely different in their outcomes and serve as an example of one way to deal effectively with adventitious viral contaminations.

The first contamination was detected at a late stage of manufacture by a general viral screening assay, which is outlined in Figure 4. In this assay, a slight positive reaction was detected after five days in cell culture in the haemadsorption (HAd) portion of the assay. The source of the HAd was identified much later as MVM by an independent murine antibody production (MAP) test. Unfortunately, by the time of this detection, the virus had apparently spread to multiple fermenters operating simultaneously. This resulted in the necessity to conduct an extensive facility decontamination and led to significant loss of product in addition to the need for time-consuming investigations and the involvement of regulatory authorities. During this episode, a full-scale analytical effort was initiated to develop methods that would be both sensitive and selective in detecting MVM. It was hoped that this effort would result in the determination of the source of the virus and that these methods could serve as an early detection system to prevent the spread of potential future contaminations. After exhaustive effort, a 324K cell culture-based infectivity assay (Fig. 5) and a

Potential Sources of Viral Contamination

- Raw materials from human or bovine origin
- Organic and inorganic media and components
- Compressed gases
- Personnel
- MCB/WCB – endogenous or adventitious agents
- Environment and facility
- Pests

Fig. 2: Potential sources of contamination from viruses that can affect large-scale cell cultures.

Minute Virus of Mice

- Autonomous parvovirus isolated in 1966
- Icosahedral virus that is not enveloped, 20 nm diameter
- Mice are the natural host. Infection is common (20%) and clinically inapparent
- Virus is excreted in urine and faeces
- Replication is dependent on factors transiently expressed during the S phase of growing host cells
- Genome is a linear single-stranded DNA molecule of ~5100 bases. The entire sequence is known
- Two strains of MVM with different specificities:
  MVM(p) – prototype strain grows in fibroblasts
  MVM(i) – immunosuppressive variant grows in T-lymphocytes

Fig. 3: Characteristics and strains of MVM.

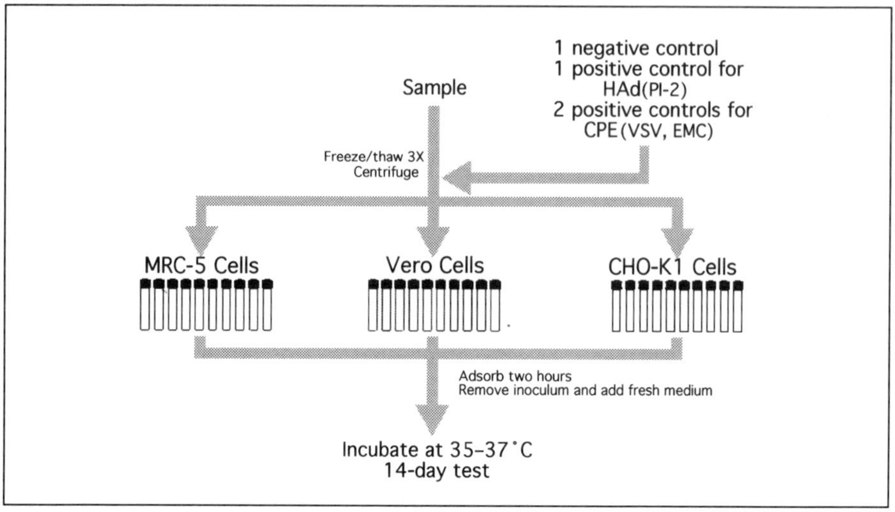

Fig. 4: Profile of the general viral screening assay. HAd = haemadsorption; PI-2 = parainfluenza virus type 2; CPE = cytopathic effect; VSV = vesicular stomatitis virus; EMC = encephalomyocarditis virus.

324 K Assay

- 324K cells are a clone of SV40 transformed human newborn kidney cells
- Inoculate ~25% confluent 324K cells and incubate
- Subculture to fresh 324K cells at days 7 and 14
- Observe periodically for CPE
- Perform final observation for CPE at day 21

Fig. 5: Profile of the 324K cell culture-based infectivity assay used successfully for detection of adventitious viruses.

polymerase chain reaction (PCR) DNA-based method were developed, optimized, and validated. These assays were then added as part of product testing in the Quality Control Department.

The time and effort spent developing these early detection assays were found to be invaluable when the second MVM infection occurred. This time the virus was detected by PCR prior to breaking containment in a single fermenter while other fermenters were in simultaneous operation. The single fermenter involved was thoroughly decontaminated while the other fermenters continued to operate. An exhaustive investigation demonstrated that the virus had not spread and there was only minimal loss of material. Based on the success of this «barrier» strategy and the early detection data, interactions with the regulatory authorities at the Food and Drug Administration (FDA) were mutually beneficial. However, this did not preclude the necessity of investigating the source of the MVM contamination. The scope of this investigation and the results are presented in Figures 6 and 7, respectively. All the evidence pointed to a non-homogeneously distributed contaminant in a raw material as the source of the MVM infection [2].

Scope of the MVM Investigation
• Internal GMP compliance audits conducted by Quality Assurance
• Examine compressed gases, water
• Examine process flow through equipment
• Quarantine raw materials, audit vendors
• Interview personnel and evaluate their training and area access
• Inspect WCB
• Evaluate pest control measures
• Characterize MVM isolate
• Examine environmental controls
• Review campaign documentation

Fig. 6: Investigation of the MVM contamination was a multi-department effort, requiring examination of numerous processes and documentation.

Results of the MVM Investigation
• Early detection of MVM combined with effective procedures for containment and decontamination prevent spread of the virus
• Facilities can continue to operate effectively
• Two different strains of MVM were isolated
• No definitive source of contamination was found either in raw materials, air, environment, or personnel
• Evidence was consistent with a non-homogeneous contaminant delivered through a raw material

Fig. 7: Results from the investigation of the MVM contamination.

Raw Material Considerations

It should be evident from the above discussion that raw materials can be a prime source of contamination by adventitious agents. However, considering the hypothesis that a viral contaminant could be non-homogeneously distributed in any raw material, the likelihood of its detection is minimal. The results of the investigation of the raw materials involved in the first MVM contamination shown in Table 1 confirm this, as in no case was a trace of MVM detected. As shown in Figure 8, the potential for contamination should also be considered when choosing the raw materials to use for a specific manufacturing process. The source of the materials is very important because the potential for contamination increases dramatically if the materials are of human or animal origin. There should be traceability of the raw material by the vendor, and the material should be amenable, if possible, to treatment by any of the methods listed that have been shown to remove or inactivate viruses (Fig. 8). The quantity of each material used in the manufacturing process is also important to know in assessing the potential for contamination from that material. The GMP operations (good manufacturing practices) in the facility itself are vital in preventing cross-contamination of multiple products or lots should an infection occur. An evaluation of even the packaging and storage areas is necessary to ensure that there are no opportunities for adventitious agents to be introduced from rodents or other pests, or through breaches in the environment.

Table 1: Raw material testing following the first MVM contamination.

Raw material	Concentration factor	324K assay	MAP test
Animal product 1	325,000	–[a]	–[a]
Animal product 2	10,240	–	–
Animal product 3	1,000	–	–
Animal product 4	1,000	–	–
Surfactant	13,000	–	–
Dextrose	1,300	–	–
Amino acid concentrate	6,500	–	–
Sodium bicarbonate	6,500	–	–
Magnesium chloride	45,500	–	–
Basal medium	4,000	–	–

a All results were negative.

Raw Material Considerations

- Source: human, animal, organic, or inorganic
- Traceability of material
- Method of production – exposure to conditions known to remove or inactivate viruses:
  - chemical inactivation
  - pasteurization
  - autoclaving
  - irradiation
- Potential for adventitious viral contamination
- Facility cleanliness: cross-contamination
- Packaging and storage areas
- Testing
- Quantity of raw materials used in the process

Fig. 8: Factors to be considered in choosing raw materials for large-scale cell culture.

Use of Analytical Methods as Barriers to Viral Contamination

The use of analytical methods to detect viruses should be considered as part of a general strategy in devising barriers against potential contamination. Such methods can be performed at the level of raw material testing or during various stages of the manufacturing process. Human source raw materials such as transferrin or albumin can contain viruses such as hepatitis (A, B, or C), human immunodeficiency virus

(HIV), or human T-cell lymphoma/leukemia virus (HTLV). Although the potential for viral contamination from these materials can be minimized by pasteurization (with validation of viral inactivation), it should be noted that the testing of such materials is only as valuable as the specificity of the assay employed. Bovine source raw materials must be tested for such viruses as bovine viral diarrhoea virus (BVDV), infectious bovine rhinotracheitis virus (IBRV), and parainfluenza virus type 3 (PI-3). Treatment of these bovine materials to prevent potential contamination can include autoclaving at 133°C or chemical treatment [in the case of bovine spongiform encephalopathy (BSE)] as well as ensuring traceability of the serum or tissues used in production.

Raw material testing by the PCR assay is not advised because of the presence of low levels of inactivated virus, which could generate false positives in some instances (Fig. 9). It cannot be overemphasized that the potential for viral contamination to be non-homogeneously distributed in a lot of raw material is very real, so there is little value to be gained by additional testing. In fact, raw material testing will only be of value under the conditions listed in Figure 10. Nonetheless, as already discussed, the PCR assay, the viral screening assay, and the 324K assay can all be successfully used at a later stage of the process in testing for the presence of viruses. These three methods complement one another, as shown in the comparison of these methods presented in Figure 11.

The general strategy outlined in this paper should help to minimize the potential for contamination of biotechnology products by adventitious agents. A summary of the concept of this barrier approach strategy is presented in Figure 12.

Considerations for the PCR Assay

- False positives
  - Specific: carryover from previous PCR amplification
  - Nonspecific: primers may anneal with genomic DNA
- False negatives due to inhibition of PCR reaction
  - Sodium dodecyl sulphate (SDS)
  - Ethylenediaminetetraacetic acid (EDTA)
  - Phenol
  - Salts [Mg]
  - pH

Fig. 9: Interpretation of results from the PCR assay must include the possibility of false positives and false negatives, which can arise from the conditions shown.

Raw Materials Testing

Testing of raw materials will not be effective in detecting viruses unless:

- contaminant is at a very high titre and homogeneously distributed
- false positives can be promptly adjudicated
- a suitably broad spectrum of viruses can be detected
- the assay has adequate sensitivity
- raw materials are not inhibitory in the assay used

Fig. 10: Testing raw materials for viruses is effective only under certain specified conditions.

CONCLUSIONS

The available data indicate that biological raw materials are the most likely source of viral contaminations in large-scale cell culture at the present time. It is imperative that a biotechnology manufacturer understand which viruses will replicate in the cell culture system(s) used because viruses that won't replicate in the cell line(s) of interest are of little concern. Specific tests must be developed for viruses that can replicate so that routine testing for them can be performed at the production stage by such methods as the PCR assay. However, it is of little value to test for viruses in the raw materials themselves. It would be virtually impossible to detect a non-homogeneous contaminant in one of the hundreds of raw materials used in the manufacture of rDNA products. The development of physical/chemical barriers to inactivate and/or remove viruses before entering the cell culture process appears to be the most effective approach to preventing viral contamination of cell cultures.

	Viral screening assay	324K assay	PCR assay
Specificity	Broad range of viruses	MVM (i) MVM (p) MVM (GNE) KRV, H1 (Rat parvoviruses)	MVM (i) MVM (p) MVM (GNE)
Sensitivity for MVM	1 $TCID_{50}$/mL (PI-2) 0.1 $TCID_{50}$/mL (Mumps) 100 $TCID_{50}$/mL (Reo-3) Unvalidated MVM	10 $TCID_{50}$/mL	0.1 $TCID_{50}$/mL 2000 genome equiv/mL
Assay run time	14 days	21 days	8 hours
Implementation schedule	Each 12K fermenter sampled preharvest	Each 12K fermenter sampled preharvest	Each 12K fermenter tested twice prior to harvest
Operations	Routine assay	Routine assay	Routine assay

Fig. 11: Comparison of the three Genentech assays used for the detection of viruses. $TCID_{50}$ = median tissue culture infectious dose; Reo-3 = reovirus type 3.

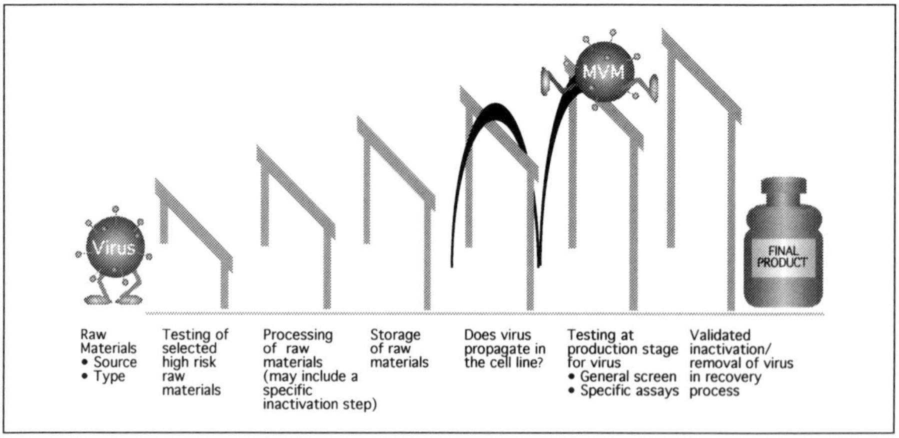

Fig. 12: Schematic of the concept of a barrier approach to preventing product contamination.

References

1 Nettleton PF, Rweyemamu MM: The association of calf serum with the contamination of BHK21 clone 13 suspension cells by a parvovirus serologically related to the Minute Virus of Mice (MVM). Arch Virol 1980;64:359-374.

2 Garnick RL: Experience with viral contamination in cell culture, in Brown F, Lubiniecki AS (eds): Viral Safety and Evaluation of Viral Clearance from Biopharmaceutical Products. Dev Biol Stand. Basel, Karger, 1996, vol 88, pp 49-56.

Dr. R.L. Garnick, Genentech, Inc., 1 DNA Way, South San Francisco, CA 94080, USA.

Brown F, Griffiths E, Horaud F, Petricciani JC (eds): Safety of Biological Products Prepared from Mammalian Cell Culture.
Dev Biol Stand. Basel, Karger, 1998, vol 93, pp 31-36

Zoonoses and Haemorrhagic Fever

B.W.J. Mahy

Division of Viral and Rickettsial Diseases, National Center for Infectious Diseases, Centers for Disease Control and Prevention, Atlanta, GA, USA

Key words: Zoonoses, haemorrhagic fever, bunyavirus, arenavirus, filovirus.

Abstract: Virus zoonoses causing haemorrhagic fever have been recognized in three major families: Arenaviridae, Bunyaviridae and Filoviridae. All are negative-stranded RNA viruses, with genomes in two segments, three segments, or non-segmented, respectively. Acquisition of haemorrhagic fever in man generally requires close contact with a vertebrate vector species, usually rodents, for the arenaviruses and bunyaviruses. In the case of filoviruses, the vector is currently unknown, but these viruses may infect monkeys, and may contaminate cell cultures prepared from them. Both bunyavirus and arenavirus haemorrhagic fevers have arisen in humans following exposure to rodents, and in the case of Hantaan, a virus causing haemorrhagic fever with renal syndrome (HFRS), there have been numerous laboratory-acquired infections among animal care workers. As the technology to differentiate virus species has improved, it has become clear that there are numerous potentially hazardous viruses capable of causing HFRS or hantavirus pulmonary syndrome (HPS) within the feral rodent population. In many cases it would be desirable to introduce screening methods for such viruses before preparing cell cultures from these rodent or simian species that will be used to prepare biological products for human use.

Introduction

Viruses causing haemorrhagic fever are mostly transmitted to humans as zoonotic infections. Some of the best known examples occur with arenaviruses, found in South America or Africa. Bunyaviruses may also cause haemorrhagic fever, or haemorrhagic fever with renal syndrome (HFRS), and these diseases are acquired from contact with rodents (HFRS: hantavirus pulmonary syndrome HPS), or sheep (Crimean-Congo haemorrhagic fever CCHF). The most dramatic examples of viral haemorrhagic fever have been observed following filovirus infection of humans or non-human primates, although the source of these zoonotic infections is currently unknown. Most recently a new zoonosis, currently correctly called equine morbillivirus, has emerged in Australia to cause at least two human deaths, as well as a larger number of deaths in horses. The disease resembles HPS in that it involves leakage of blood plasma into the lungs, but the causative virus is not yet well characterized.

ARENAVIRUSES

Table 1 lists the currently known arenaviruses and their rodent vector species. They can be divided geographically and taxonomically into two antigenic complexes, Old World and New World arenaviruses. A number of newly recognized New World arenaviruses are capable of causing haemorrhagic fever, but so far the number of cases described is small. The arenavirus genome consists of two segments of negative-strand RNA. There may be some similarities observed by genome RNA sequence analysis between those viruses capable of causing severe haemorrhagic fever on the one hand and on the other those that have not been associated with haemorrhagic disease.

Table 1: Arenaviruses.

Old world	**Rodent host**	**Year recognized**
Ippy	*Arricanthus spp*	1985
LCM	*Mus spp*	1933
*Lassa	*Mastomys spp*	1969
Mobala	*Praomys spp*	1981
Mopeia	*Mastomys natalensis*	1977
New world	**Rodent host**	**Year recognized**
Amapari	*Oryzomys goedi* *Neacomys guianae*	1964
Flexal	*Oryzomys spp.*	1975
*Guanarito	*Zygodontomys brevicauda*	1990
*Junin	*Calomys musculinus*	1958
Latino	*Calomys callosus*	1965
*Machupo	*Calomys callosus*	1963
Oliveros	*Bolomys obscurus*	1990
Parana	*Oryzomys buccinatus*	1965
Pichinde	*Oryzomys albigularis*	1965
Pirital	*Sigmodon alstoni*	1995
*Sabia	Not known	1993
Tacaribe	*Artibeus bats*	1956
Tamiami	*Sigmodon hispidus*	1964
Whitewater Arroyo	*Neotoma albigula*	1995

* Indicates virus associated with haemorrhagic fevers in humans.

A considerable potential exists for arenavirus contamination of cell cultures, and this has been reviewed [1]. Most rodent colonies are now routinely screened for the possible presence of lymphocytic choriomeningitis virus infection, which can spread rapidly and has caused a number of human infections.

Bunyaviruses

Members of the family *Bunyaviridae* have a genome consisting of three segments of negative-strand RNA. There are five recognized genera, but only two (*Hantavirus* and *Nairovirus*) contain virus species that cause haemorrhagic fever. Hantaviruses are present in various rodent species in which they establish a usually silent, persistent infection. Humans generally become infected from contact with rodent excreta, but in at least one instance infection occurred in laboratory workers who had handled immunocytomas propagated in rats, without exposure to the animals themselves [2]. It is therefore advisable to screen cell lines or hybridomas that have been passed in rodents for the possible presence of hantavirus.

Since the discovery of Sin Nombre virus and related viruses causing hantavirus pulmonary syndrome (HPS), a severe disease in humans with more than 50% mortality [3], it has become clear that infections with hantaviruses are widespread in rodents of the subfamily *Sigmodontinae*. These include Sin Nombre virus in the deer mouse (*Peromyscus maniculatus*), a related virus in the white-footed mouse (*Peromyscus leucopus*), Black Creek Canal virus in the cotton rat (*Sigmodon hispidus*), and Bayou virus in the rice rat (*Oryzomys palustris*). All these viruses have caused serious HPS infections in humans, many of them fatal. In addition, evidence of hantavirus infection has been obtained from genetic sequence analysis of tissues of several species of voles in the subfamily *Arvicolinae*. It will be important in the future to develop appropriate screening methods for cells and cell lines derived from such species, similar to those applied in the case of Hantaan and Seoul viruses [4].

The genus *Nairovirus* includes one important member causing severe haemorrhagic fever, CCHF virus. Although the virus is maintained in nature in *Hyalomma* ticks, which may infect humans by biting, large animals such as sheep provide a direct source of transmission to veterinarians and slaughter-house workers handling their infected tissues, leading to occasional disease outbreaks with up to 50% mortality.

Filoviruses

There are two recognized species in the family *Filoviridae*, Marburg virus and Ebola virus. Their genome RNA is a single continuous negative strand. The sudden appearance of Marburg virus in 1967, causing outbreaks of haemorrhagic fever among laboratory workers preparing kidney cell cultures of African green monkey (*Cercopithecus aethiops*) in Germany and Yugoslavia, has been well described [5]. Since the original episode, isolated cases of Marburg haemorrhagic fever have occurred in Africa, the last in 1987 (Table 2). Epidemics of severe haemorrhagic fever due to Ebola virus occurred between 1976 and 1979 in Zaire and Sudan, then the virus disappeared for some ten years. No natural reservoir for this virus has

Table 2: Episodes of Marburg virus infection.

Year	Location	Number of human cases	Percent mortality
1967	Germany and Yugoslavia	31	23
1976	Zimbabwe	3	33
1980	Kenya	2	50
1987	Kenya	1	100

been found. However, in late November 1989, a quarantine facility in Reston, Virginia, noticed an unusually large number of sick cynomolgus macaques (*Macaca fascicularis*) among a batch of animals recently imported from the Philippines. Infection with simian haemorrhagic fever (SHF) virus was suspected, and tissues from sick or dead animals were sent to the U.S. Army Medical Research Institute of Infectious Diseases (USAMRIID) at Fort Detrick, Maryland. Although SHF virus was found, tissue material from one monkey, when cultivated on MA-104 cells, resulted in the growth of filovirus-like particles seen upon EM examination. The filovirus was not Marburg, as first suspected, but an Ebola-related virus based on its reaction with specific monoclonal and polyclonal antisera [6]. During 1990 additional filovirus-infected monkeys were detected in Pennsylvania, Texas, and again in Virginia and the virus was isolated and named Ebola Reston virus [7, 8]. It was clearly distinct antigenically and genetically from the Ebola Zaire and Ebola Sudan viruses. There also appear to be differences in pathogenicity for humans, since four persons became infected with Ebola Reston at the Virginia facility but none developed any disease [7]. However, Ebola Reston virus is highly pathogenic for non-human primates in both natural and experimental infections. Further outbreaks of Ebola Reston occurred in Siena, Italy in 1992 and in Alice, Texas in 1996. In every case the infected monkeys appear to have been exported from one particular company in Manila, Philippines. Extensive ecological investigations have so far failed to identify the source of these infections, and the possibility remains that further Ebola Reston disease outbreaks may occur in non-human primates received from the Philippines. However, it became clear during the 1996 outbreak in monkeys in Alice, Texas, that the current guidelines which require one month's quarantine of all non-human primates newly introduced into the USA are effective in detecting and preventing the spread of filovirus infection.

In the past two years, four further episodes of human infection with Ebola have occurred (Table 3). The most serious, centering on a hospital in Kikwit, Zaire, involved a charcoal burner as the index case in January 1995, from whom many unrecognized secondary infections with haemorrhagic fever occurred over the subsequent five month period. It was not until May that the causative virus was recognized as Ebola Zaire, and found by phylogenetic analysis to be very closely related to the Zaire strain isolated 19 years earlier [9]. Application of barrier nursing techniques and health education of the community brought the epidemic to a close within two months, and fortunately it was contained without spread outside the country. An intensive investigation is under way to find a possible reservoir from which the index case became infected.

Table 3: Episodes of Ebola virus infection.

Year	Location	Number of human cases	Percent mortality
1976	Northern Zaire	318	88
1976	Southern Sudan	284	53
1976	England (laboratory infection)	1	0
1977	Southern Zaire	1	100
1979	Southern Sudan	34	65
1989	Reston, VA, USA	4	0
1992	Siena, Italy	0	0
1994	Ivory Coast	1	0
1994	Gabon	49	65
1995	Southern Zaire	316	77
1996	Gabon	37	57
1996	Alice, TX, USA	0	0

The other three recent episodes in the Ivory Coast and Gabon have all involved contact between humans and sick or dead chimpanzees. The single infection in the Ivory Coast occurred in a person who performed an autopsy on a dead chimpanzee, and from the patient's blood an Ebola virus was isolated which is genetically distinct from the known subtypes Ebola Zaire, Ebola Sudan or Ebola Reston, and is now considered to be a new subtype, Ebola Ivory Coast. The 1996 episode in Gabon resulted in the death of 21 of 37 persons who handled the meat of a dead chimpanzee before it was cooked for consumption. The Gabon virus belongs to the Ebola Zaire virus subtype.

Acknowledgements

I wish to thank numerous colleagues in the Special Pathogens Branch, Division of Viral and Rickettsial Diseases for assistance and information, with special thanks to Michael Bowen, Tom Ksiazek, Stuart Nichol, C.J. Peters, Pierre Rollin and Tony Sanchez.

References

1 Mahy BWJ, Dykewicz C, Fisher-Hoch S, Ostroff S, Tipple M, Sanchez A: Virus zoonoses and their potential for contamination of cell cultures, in Horaud F, Brown F (eds): Virological Aspects of the Safety of Biological Products. Dev Biol Stand, Basel, Karger, 1991, vol 75, 183-189.

2 Lloyd G, Jones N: Infection of laboratory workers with hantavirus acquired from immunocytomas propagated in laboratory rats. J Infect 1986;12:117-125.

3 Khan AS, Khabbaz RF, Armstrong LR, Holman RC, Bauer SP, Graber J, Strine T, Miller G, Reef S, Tappero J, Rollin PE, Nichol ST, Zaki SR, Bryan RT, Chapman LE, Peters CJ, Ksiazek TG: Hantavirus pulmonary syndrome: The first 100 US cases. J Infect Dis 1996;173:1297-1303.

4 LeDuc JW, Smith GA, Macy M, Hay RJ: Certified cell lines of rat origin appear free of infection with hantaviruses. J Infect Dis 1985;152:1082-1083.

5 Martini GA, Siegert R (eds): Marburg virus disease. New York, Springer-Verlag, 1971.

6 CDC Update: Ebola-related filovirus infection in non-human primates and interim guidelines for handling non-human primates during transit and quarantine. Morb Mortal Weekly Rpt 1990; 29:29-30.

7 CDC Update: Filovirus infections among persons with occupational exposure to non-human primates. Morb Mortal Weekly Rpt 1990;39:266-267, 273.

8 Jahrling, PB, Giesbert TW, Dalgard DW, Johnson ED, Ksiazek TG, Hall WC, Peters CJ: Preliminary report: Isolation of Ebola virus from monkeys imported to USA. Lancet 1990;335:502-505.

9 Sanchez A, Trappier SG, Mahy BWJ, Peters CJ, Nichol ST: The virion glycoproteins of Ebola viruses are encoded in two reading frames and are expressed through transcriptional editing. Proc Natl Acad Sci USA 1996;93:3602-3607.

Dr. B.W.J. Mahy, Division of Viral and Rickettsial Diseases, National Center for Infectious Diseases, Centers for Disease Control and Prevention, 1600 Clifton Road, N.E., Mailstop A30, Atlanta, GA 30333, USA.

Brown F, Griffiths E, Horaud F, Petricciani JC (eds): Safety of Biological Products Prepared from Mammalian Cell Culture.
Dev Biol Stand. Basel, Karger, 1998, vol 93, pp 37-44

Overview of Biological Effects of Addition of DNA Molecules to Cells*

[†] *H.M. Temin*

McArdle Laboratory, University of Wisconsin, Madison, WI, USA

Key words: Retrovirus, transfection, degradation, adsorption, uptake, ligation, proto-oncogene, inactivation, activation, tumour suppressor gene.

Abstract: Injected DNA proceeds with certain probabilities through the following steps: degradation by serum nucleases, adsorption to cells, uptake into cells, ligation to other DNA, mutation, expression of unintegrated DNA, integration, expression of integrated DNA, and activation of or inactivation of cellular genes. The maximal probability per DNA molecule of each of these steps is estimated based on experimental results in cell culture with transfection of DNA and with infection by retroviruses. A maximum cumulative probability of having a harmful effects is calculated to be less than 10^{-16} to 10^{-19} per DNA molecule from a cell without activated proto-oncogenes or active viral oncogenes. The most frequent harmful effects considered are inactivation of a tumour suppressor gene and activation of a proto-oncogene. Such inactivation and activation in a cell that could give rise to cancer would increase the age-standardized incidence of cancer by a small amount. The amount of increase would differ among individuals depending upon their genotypes and their environments. Thus, the magnitude of the increase will depend upon the frequency of more sensitive individuals. The probability of an increased incidence of cancer as a possible effect of the vaccination should be compared to the number of DNA molecules to be injected per person and to the protective effects of successful HBV vaccine.

INTRODUCTION

The purpose of this paper is to calculate the probability per DNA molecule that there will be a harmful effect of a DNA molecule injected into a person. Injected DNA proceeds with certain probabilities through the following steps: degradation by serum nucleases, adsorption to cells, uptake into cells, ligation to other DNA, mutation, expression of unintegrated DNA, integration, expression of integrated DNA, and activation or inactivation of cellular genes. Although gene inactivation can be lethal to a cell, effects on the entire organism are seen only when the cellular genetic changes result in or contribute to a selectable phenotype for the altered cell that is deleterious to the organism, that is, the added DNA results in increasing the

† Deceased.
* Reprinted by permission of Wiley-Liss, Inc., a Division of John Wiley & Sons, Inc., Journal of Medical Virology, 1990, vol 31, pp 13-17.

probability of cancer. Thus, only activation of proto-oncogenes and inactivation of tumour-suppressor genes in cells that can give rise to cancer are important consequences of the addition to cells of exogenous DNA from cells without activated proto-oncogenes or active viral oncogenes.

The probability per DNA molecule of each of the steps – degradation by serum nucleases, adsorption to cells, uptake into cells, ligation to other DNA, mutation, expression of unintegrated DNA, integration, expression of integrated DNA, and activation of or inactivation of cellular genes – can be estimated based on experimental results in cell culture with transfection of DNA and with infection with retroviruses. I shall assume that the results in cell culture can be transferred to humans and that the sizes of the DNA molecules do not affect the probabilities. Since the cells in culture are used because of their sensitivity, these probabilities are higher than for most cells in the body. A maximum cumulative probability of having a harmful effect is calculated to be less than 10^{-16} to 10^{-19} per DNA molecule from a cell without activated proto-oncogenes or active viral oncogenes. The most frequent harmful effects considered are inactivation of a tumour suppressor gene and activation of a proto-oncogene. Such inactivation or activation in a cell that could give rise to cancer would increase the age-standardized incidence of cancer by a small amount. The amount of increase would differ among individuals depending upon their genotypes and their environments. Thus, the magnitude of the increase will depend upon the frequency of more sensitive individuals. The probability of an increased incidence of cancer should be compared to the number of DNA molecules to be injected per person and to the protective effects of a successful HBV vaccine.

Fate of Injected DNA Molecules

The amount of degradation of injected DNA before adsorption to cells depends upon the type and concentration of nucleases in blood. However, since there is very efficient ligation of DNA taken up into cells (discussed below), unless the DNA is degraded to small oligonucleotides or nucleotides, the injected DNA molecules will be rejoined and hence still capable of integration (see below).

The efficiency of entrance of the injected DNA into cells depends upon the efficiency of adsorption to cells and the efficiency of uptake into cells. Over 50% of many different types of cells in culture take up DNA after appropriate types of treatments and, if the DNA has a promoter, express it transiently. For example, 50-75% of human peripheral blood lymphocytes expressed a plasmid after electroporation [1]. Furthermore, stable transfection of over 50% of the cells in mammalian cell lines can be achived [2]. However, there is a very low efficiency per DNA molecule in both of these cases.

Kawai and Nishizawa [3] showed with transfection of chicken embryo fibroblasts in culture that adding Polybrene, wich increases adsorption, increased transfection efficiency by over 100-fold, and treatment with DMSO, which enhances uptake, also increased transfection efficiency by the same factor. However, even at optimal conditions with very sensitive cells and with DNA with a strong promoter, the efficiency of entrance, as measured by expression and assuming expression of all molecules that entered, was only 2×10^{-7} per DNA molecule. (The efficiency of entrance is actually greater if the efficiency of expression is less than 100%). Furthermore the cells in a person most likely to encounter injected DNA are lymphocytes which are very difficult to transfect.

Fate of Intracellular DNA Molecules

Ligation and mutation

When DNA enters avian or mammalian cells in culture by transfection or injection, it both is ligated with high efficiency and suffers mutations. For example, Miller and Temin [4] found that the efficiency of recombination of transfected retrovirus DNA in chicken and dog cells was about 100%. Wake et al [5] found that after transfection of monkey cells there was nearly 100% joining; in addition, there was one break induced per 5-15 kbp. Dougherty and Temin [6] found a rate of deletion mutations of about 20% after transfection of dog cells in culture with retrovirus vectors. Kopchick and Stacey [7] found that after transfection of quail tumour cells in culture there was efficient ligation, but most, over 95%, of the ligated molecules had mutations. Thus, DNA molecules taken up into cells will be ligated together and will suffer mutations, primarily deletions.

Expression

As discussed above, unintegrated intracellular DNA with a strong promoter can be expressed in a high percentage of cells. After transfection or elctroporation, over 50% of cells can show evidence of expression of input DNA. Since there are many DNA molecules added per cell while under optimal conditions the efficiency of entrance into cells is about 10^{-7}, it is hard to tell the efficiency of expression per intracellular DNA molecule. Using a retrovirus vector system that allowed comparison of expression of intracellular cDNA with that of retrovirus vector DNA, Dornburg and Temin [8, 9] established that a high percentage of unintegrated cDNA genes with a promoter were expressed. However, these DNA molecules were formed after infection, not transfection, and so might have a different intracellular location. In any case, it is hard to imagine any harmful effect on the organism from expression of unintegrated contaminating DNA molecules. (Of course, this generalization does not apply to virus DNA).

Integration of molecules that have entered cells

The efficiency of integration of intracellular DNA in cells in culture is dependent on the cell type. For example, stable transfection of chicken fibroblasts in culture is rare. However, injection of *src* DNA into chickens results in tumours at an efficiency of one tumour per 3×10^{11} injected DNA molecules [10].

However, in cell types where stable integration does occur, the efficiency appears high. For example, Dornburg and Temin [8, 9], using a retrovirus vector system for studying the fate of cDNA sequences, showed that about 20% of intracellular DNA molecules with no special sequences promoting integration were stably integrated in dog cells in culture. However, the frequency of stable integration per molecule after transfection of cells in culture is usually much lower. The discrepancy between these two types of experiments may reflect the different routes of entrance into cells by infection and transfection or the difference in assays for integration and expression. However, the higher figure indicates that integration of intracellular DNA molecules with no special mechanisms for integration can be very efficient under some circumstances. It also suggests that these special circumstances may not exist for contaminating injected DNA.

One study [11] suggests that transfected DNA may integrate preferentially in repetitive DNA. If this were generally true, it would reduce the inactivating effects discussed below. However, Dornberg and Temin [12] found that two of three integrants of cDNA genes were in unique sequences suggesting that the Kato et al [11] results are not general.

Effect of Integrated DNA Molecules

Expression

Integrated DNA can be expressed from an internal promoter or from a promoter in the neighboring cellular DNA. The frequency of expression depends upon the nature of the cells and the internal promoters. For example, integrated retrovirus DNA is usually not expressed in embryonal carcinoma cells as a result of the presence of certain control sequences (for example, [13]). A harmful effect to the organism of this expression would only be possible if an intact gene for a toxin, an activated proto-oncogene, or an oncogene was present in contaminating DNA.

Activation of proto-oncogenes

Expression from a promoter in the integrated DNA or the presence of an enhancer sequence in this DNA can lead to transcriptional activation of a nearby gene. This activation would be analogous to insertional activation by weakly oncogenic retroviruses. When the activated gene is a proto-oncogene, this insertional activation can represent one step towars carcinogenesis. The frequency of insertional activation of proto-oncogenes by integrated DNA depends upon the number of cellular proto-oncogenes whose activation has an effect on a particular type of cell and the frequency of the presence of enhancers or promoters in the integrated DNA. The frequency of activation of a particular proto-oncogene by retrovirus integration is probably less than 10^{-5} per integration, assuming that the target for insertional activation is about 30 kbp, assumed to be the size of a typical gene (there are 3×10^9 bp/haploid genome). Since there are probably about 100 proto-oncogenes per cell, the probability of a random retrovirus provirus causing activation of a proto-oncogene is less than 10^{-3} per cell.

In addition, two papers [14, 15] propose that some specific short pieces of transfected DNA not containing oncogenes can cause transformation of immortal cells in culture at an efficiency of 1-10 per µg per 10^6 cells. Other authors [16] proposed that such stable integration of transfected DNA not coding for oncogenes promotes chromosomal rearrangements, especially trisomy in their experiments. This would be another mechanism of gene activation, perhaps not dependent upon the presence of promoters or enhancers in the integrated DNA.

Since enhancers, promoters, or sequences affecting chromosomal rearrangements are clustered along the chromosomes and only represent a small fraction of the total cellular genome, they would only be in approximately 1% of DNA molecules (depending upon the source of DNA and the size of the DNA molecules), and most molecules of cellular DNA would not contain them. Thus activation would only be seen with perhaps 1 out of 10^2 DNA molecules integrated near a proto-oncogene. Since oncogenes are cell specific, there are probably less than ten proto-oncogenes whose activation affects any particular cell type. Therefore, the probability of a random molecule of integrated DNA causing activation of a proto-oncogene affecting a particular cell is less than 10^{-6} per cell (assuming the same target size as for retroviral insertional activation). This activation would only have an effect if that cell had the potential to give rise to cancer.

This discussion has assumed that the contaminating DNA comes from cells without activated proto-oncogenes or active viral oncogenes. I now consider the effects of contamination by DNA from cells with activated proto-oncogenes. Assuming that there is one such gene per cell there would be a 10^{-5} chance that a random DNA molecule contains the activated proto-oncogene and a 10-30% chance that its specificity is appropriate for a particular cell. The probability of tumour formation in sensitive young chickens by DNA from a Rous sarcoma virus is 3×10^{-12} per molecule [10], giving the probability of integrating a small intact molecule. Thus, the maximum probability of any biological effect of contaminating DNA from a cell with an active proto-oncogene is about 10^{-17}.

If the contaminating DNA was from a cell with an active viral oncogene, the probability of inducing a biological effect would be similar, but the effect on the age-standardized incidence of cancer might be greater. However, since human cells are more resistant to transformation than the chicken or rodent cells used for assay of viral oncogenes [17] and the experiments involved pure populations of cloned oncogene DNA, these estimates are almost certainly much too high.

Gene inactivation

Inactivation as a result of integration is also a major effect of integration of contaminating DNA. Such inactivation can be the result of the interruption of coding sequences or the presence of control sequences that interfere with expression of the surrounding gene. For example, integration of retrovirus DNA can lead to inactivation. Varmus et al [18] showed that integration of a murine leukemia virus provirus inactivated a resident Rous sarcoma virus provirus. Jaenisch [19] estimated that 5-6% of retroviruses integrated into early mouse embryos cause mutations. For example, insertion of retrovirus DNA into the first intron of the (alpha)1(I) collagen gene leads to an embryonic lethal mutation in mice [20]. Similarly, Lock et al [21] found 5% recessive lethals after acquisition of proviruses by infection of mouse embryos. These results indicate that if integration is random, the target for retroviral inactivation is about 10^{-8} bp ($0.05 \times 3 \times 10^9$ bp). If there 10^3 essential genes, the target size per gene is similar to that for retroviral activation. Furthermore, there is some evidence suggesting that retrovirus integration is more likely in transcriptionally active chromatin [22, 23]. Similarly, integration of transposon DNA frequently leads to inactivations of genes [24]. For example, in *Drosophila* many spontaneous mutations are the result of transposon insertion.

Another way to look at insertional inactivation is to consider the efficiency of inactivation of a particular gene as a result of retrovirus integration. Some researchers [25, 26] measured inactivation of the *hprt* gene in embryonal carcinoma and stem cells in culture after retrovirus infection and found a frequency of about 1 inactivation per 10^8 proviruses or in 1 out of 10^7 multiply infected cells, respectively. The large differences between these numbers and the frequency of lethal mutations after retrovirus integration in early embryos might reflect a difference in the cells or a special feature of the *hprt* gene. However, the *hprt* gene is 33 kbp with 1.307 bp of coding sequences spread over nine exons [27], and the inactivating proviruses were integrated in introns [25-26].

Retrovirus DNA contains in 10 kbp at least two strong enhancer-promoters and two polyadenylation signals. Thus, integration of contaminating cellular DNA molecules would be much less efficient in causing inactivation. Usually only integration in coding sequences would result in activation.

There is some evidence about the effects of integration of DNA into the human genome. The human genome contains well over 500.000 integrants, especially of Alu sequences [28]. The integrants present in DNA of present day organisms seem to be selectively neutral in their effect, except possibly for increasing the frequency of some chromosomal deletions.

Worst Case Scenario

Accepting these data, what is the probability that a contaminating piece of DNA from a cell without activated proto-oncogenes or active viral oncogenes would have any effect on a cell and on the organism of which it is a part? Gene inactivation could kill a cell, but that would have no effect on the whole organism. Inactivation of a tumour suppressor gene, such as the retinoblastoma gene which is about 190 kbp with about 2.5 kbp of coding sequence [29], or activation of a proto-oncogene, would in the appropriate cell represent one step in carcinogenesis. The probability of integration of a random piece of DNA into a tumour suppressor gene is about 10^{-5} (assuming a target size like that for mouse embryonic lethals or 30 kbp) or about 10^{-8} (assuming a target size like that of the *hprt* gene). There are probably fewer than 100 tumour suppressor genes per cell, so the probability is 10^{-3} to 10^{-6} per DNA molecule of integration into some tumour suppressor gene. The probability that a molecule of DNA integrated into a tumour suppressor gene will inactivate that gene is probably well less than 10^{-2}, since only integration into a coding sequence would be inactivating. (This is different from the effect of retrovirus DNA which can also inactivate when integrated in non-coding sequences.) The probability of integration of a molecule of intracellular DNA can be as high as 10^{-1}. Thus an intracellular DNA molecule has at most a 10^{-6} to 10^{-9} chance of inactivating a tumour suppressor gene. (This number is lower if some tumour suppressor genes are cell or tissue specific.) The probability of activating a proto-oncogene affecting a particular cell appears to be similar.

The probability of a molecule of DNA surviving nucleases and entering a cell is certainly lower than 10^{-10} per DNA molecule. (Recall that under optimized conditions of transfection, the probability of entrance of a DNA molecule is 2×10^{-7}). Thus, per injected DNA molecule there is less than a 10^{-16} to 10^{-19} probability of inactivating a tumour suppressor gene.

What is the effect of inactivating a tumour suppressor gene in a random cell on the incidence of cancer? Since only a fraction of the cells in the body can give rise to cancer, this inactivation can only have an effect in this fraction of cells. Carcinogenesis is a multi-step process with the number of steps ranging from a few for certain leukemias and childhood tumours to perhaps over ten for colon cancer [30-36].

Inactivation of an appropriate tumour suppressor gene in a cell capable of growing into a cancer would increase the age-standardized incidence of cancer by some amount. In large population of people there almost certainly will be some whose genotype or environment makes them more sensitive to this effect. In such persons, the incidence of cancer could be increased to a greater extent. Thus, the overall magnitude of the increase will depend upon the frequency of more sensitive individuals in the vaccinated population.

However, this possible increase in the incidence of cancer has to be balanced against the decrease in incidence of cancer as a result a successful hepatitis B virus (HBV) vaccine. Since HBV causes cancer in a known percentage of infected people and it can be calculated what percentage of those to be vaccinated would have otherwise become infected by HBV, a riskbenefit ratio can be determined. This ratio then can be used in deciding whether or not use the vaccine.

Acknowledgements

I thank John Cairns, Wei-Shau Hu, Gary Pulsenilli, and Shiaolan Yang for helpful comments on this manuscript. The research in my laboratory is supported by Public Health Service research grants CA-22443 and CA-07175 from the National Cancer Institute. I am an American Cancer Society research professor.

References

1 Cann AJ, Koyanagi Y, Chen ISY: High efficiency transfection of primary human lymphocytes and studies of gene expression. Oncogene 1988;3:123-128.

2 Chen C, Okayama H: High-efficiency transformation of mammalian cells by plasmid DNA. Molecular and Cellular Biology 1987;7:2745-2752.

3 Kawai S, Nishizawa M: New procedure for DNA transfection with polycation and dimethylsulfoxide. Molecular and Cellular Biology 1984;4:1172-1174.

4 Miller CK, Temin HM: High-efficiency ligation and recombination of DNA fragments by vertebrate cells. Science 1983;220:606-609.

5 Wake CT, Gudewicz T, Porter T, White A, Wilson JF: How damaged is the biologically active subpopulation of transfected DNA? Molecular and Cellular Biology 1984;4:387-398.

6 Dougherty JP, Temin HM: High mutation rate of a spleen necrosis virus-based retrovirus vector. Molecular and Cellular Biology 1986;6:4387-4395.

7 Kopchick JJ, Stacey DW: Differences in intracellular DNA ligation after microinjection and transfection. Molecular and Cellular Biology 1984;4:240-246.

8 Dornburg R, Temin HM: Retroviral vector system for the study of cDNA gene formation. Molecular and Cellular Biology 1988;8:2328-2334.

9 Dornburg R, Temin HM: Presence of a retroviral encapsidation sequence in nonretroviral RNA increases the efficiency of formation of cDNA genes. Journal of Virology 1990b;64:886-889.

10 Fung Y-KT, Crittenden LB, Fadly AM, Kung H-J: Tumor incidence by direct injection of cloned v-*src* DNA into chickens. Proceedings of the National Academy of Sciences of the United States of America 1983;80:353-357.

11 Kato S, Anderson RA, Camerini-Otero RD: Foreign DNA introduced by calcium phosphate is integrated into repetitive DNA elements of the mouse L genome. Molecular and Cellular Biology 1986;6:1787-1795.

12 Dornburg R, Temin HM: cDNA genes formed after infection with retroviral vector paticles lack the hallmarks of natural processed pseudogenes. Molecular and Cellular Biology 1990a;10:68-74.

13 Loth TP, Sievert LL, Scott RW: Negative regulation of retrovirus expression in embryonal carcinoma cells mediated by an intragenic domain. Journal of Virology 1988;62:4086-4095.

14 Lau CC, Gadi IK, Kalvonjian S, Anisowicz A, Sager R: Plasmid-induced «hit-and-run» tumorigenesis in Chinese hamster embryo fibroblasts (CHEF) cells. Proceedings of the National Academy of Sciences of the United States of America 1985;82:2839-2843.

15 Galloway DA, Nelson JA, McDougall JK: Small fragments of herpesvirus DNA with transforming activity contain insertion sequence-like structures. Proceedings of the National Academy of Sciences of the United States of America 1984;81:4736-4740.

16 Steneman G, Delmore EO, Lau CG, Sager R: Transfection with plasmid pSV2gptEJ induces chromosome rearrangements in CHEF cells. Proceedings of the National Academy of Sciences of the United States of America 1987;84:184-188.

17 Sager R: Genetic suppression of tumor formation: A new frontier in cancer research. Cancer Research 1986;46:1573-1580.

18 Varmus HE, Quintrell N, Oritz S: Retroviruses as mutagens: insertion and excision of a non-transforming provirus alter expression of a resident transforming provirus. Cell 1981;25:23-36.

19 Jaenisch R: Transgenic animals. Science 1988;240:1468-1474.

20 Harbers K, Kuehn M, Delius H, Jaenisch R: Insertion of retrovirus into the first intron of (alpha) (I) collagen gene leads to embryonic lethal mutation in mice. Proceedings of the National Academy of Sciences of the United States of America 1984;81:1504-1508.

21 Lock LF, Keshet E, Gilbert DJ, Jenkins NA, Copeland NG: Studies of the mechanism of spontaneous germline ecotropic provirus acquisition in mice. EMBO Journal 1988;7:4169-4177.

22 Vijaya S, Steffen LL, Robinson LL: Acceptor sites for retroviral integrations map near DNase I-hypersensitive sites in chromatin. Journal of Virology 1986;60:683-692.

23 Rohdewohld H, Weiher H, Reik W, Jaenisch R, Breindl M: Retrovirus integration and chromatin structure: Moloney murine leukemia proviral integration sites map near DNase I-hypersensitives sites. Journal Virology 1987;61:336-343.

24 Shapiro JA (ed): Mobile Genetic Elements. New York, Academic Press, 1983.

25 King W, Patel MD, Lobel LI, Goff SP, Nguyen-Huu MC: Insertion mutagenesis of embryonal carcinoma cells by retroviruses. Science 1985;228:554-558.

26 Kuehn MR, Bradley A, Robertson EJ, Evans MJ: A potential animal model for Lesch-Nyhan syndrome through introduction of HPRT mutations into mice. 1987;326:295-298.

27 Melton DW, Konecki DS, Brennard J, Caskey CT: Structure, expression, and mutation of the hypoxanthinephosphoribosyl-transferase gene. Proceedings of the National Academy of Sciences of the United States of America 1984;81:2147-2151.

28 Temin HM: Reverse transcription in the eukaryotic genome: Retroviruses, pararetroviruses, retrotransposons, and retrotranscripts. Molecular Biology and Evolution 1985;2:455-468.

29 Lee W-H, Bookstein R, Hong F, Young L-J, Shew J-Y, Lee E: Human retinoblastoma susceptibility gene: Cloning, identification, and sequence. Science 1987;235:1394-1399.

30 Peto R: Epidemiology, multistage models, and short-term mutagenicity tests, in Hiatt HH, Watson JD, Winsten JA (eds): Origins of Human Cancer. Vol. C Human Risk Assessment. Cold Spring Harbor Conferences on Cell Proliferation. New York, Cold Spring Harbor Laboratory, 1977, vol 4, pp 1403-1428.

31 Foulds L: The natural history of cancer. Journal of Chronic Diseases 1958;8:2-37.

32 Farber E: Cellular biochemistry of the stepwise development of cancer with chemicals: GHA Clowes Memorial Lecture. Cancer Research 1984;44:5463-5474.

33 Suarez HG, Nardeux PC, Andeol Y, Sarasin A: Multiple activated oncogenes in human tumors. Oncogene 1987;1:201-207.

34 Liu E, Dollbaum C, Scott G, Rochlitz C, Benz C, Smith HS: Molecular lesions involved in the progression of a human breast cancer. Oncogene 1988;3:323-327.

35 Vogelstein B, Fearon ER, Kern SE, Hamilton SR, Preisinger AC, Nakamura Y, White R: Allelotype of colorectal carcinomas. Science 1989;244:207-211.

36 Baker SJ, Fearon ER, Nigro JM, Hamilton SR, Preisinger AC, Jessup JM, Van Tuinen P, Ledbetter DH, Barker DF, Nakamura Y, White R, Vogelstein B: Chromosome 17 deletions and p53 gene mutations in colorectal carcinomas. Science 1989;244:217-221.

Brown F, Griffiths E, Horaud F, Petricciani JC (eds): Safety of Biological Products Prepared from Mammalian Cell Culture.
Dev Biol Stand. Basel, Karger, 1998, vol 93, pp 45-56

Risk Potential of the Chromosomal Insertion of Foreign DNA*

R. Kurth

Paul-Ehrlich-Institut, Langen, Germany

INTRODUCTION

Foreign DNA enters living organisms in three ways: during fertilization, in food and in infections. Obviously, the uptake of DNA in sperm cells to fertilize oocytes is a tightly regulated process. Exposure to DNA in food is unavoidable and therefore of no harmful consequence, except in very rare cases of allergic reactions. DNA or RNA of invading micro-organisms are very efficiently introduced into the infected host but usually persist only transiently. Exceptions are made by a few virus families, the member strains of which may persist life-long after infection like the retroviruses or the herpes simplex viruses.

In modern times, DNA or RNA may also enter animals or humans via the doctors's needle, for example, in the context of gene therapy clinical trials or nucleic acid vaccination. Safety has been a consideration in all of these gene therapy clinical trials or DNA vaccination experiments. Among these, the risks of inducing tumours, anti-DNA antibodies or other autoimmune disorders figure prominently.

The risks of inducing anti-DNA antibodies or immunological disorders such as specific tolerance, anaphylaxis, auto- or hyperimmunity are more hypothetical. It is exceedingly difficult to raise antibodies to native double-stranded DNA, although easier against denatured ssDNA. The other disorders mentioned above have not yet been described following DNA vaccination, but the observation period may still be too short. The frequency of their induction, if it occurs at all, should be lower than that seen with the live virus vaccines in clinical use.

The risk of malignant transformation of cells that have taken up injected or transfected nucleic acids is not exclusively theoretical. For example, injection of retroviral oncogene DNA into a susceptible host has led to dose-dependent tumour development [1]. Furthermore, foreign DNA may become chromosomally integrated and the integration event may result in protooncogene activation or tumour suppressor gene inactivation [2, 3]. Corresponding risk assessment is the subject of this article.

---

* Reprinted by permission of the author. Published in the Annals of the New York Academy of Sciences, 1995, vol 772, pp 140-151.

When discussing risks, one cannot overlook potential benefits. Obviously, the risk/benefit ratio gene therapy is very different from that in DNA vaccination. In the former, patients are normally treated who suffer from very serious diseases and who possess a very poor prognosis. In contrast, vaccinees are usually young and healthy. Thus, the higher relative risk in nucleic acid vaccination (because vaccinees are not patients) is an important aspect in the ongoing discussions about safety.

Natural DNA Transposition in the Eukaryotic Genome

Mother Nature provides us with numerous examples of de novo DNA integration and its consquences, providing a background for the initial evaluation of safety issues related to vaccine DNA with the potential to integrate [4].

Transposed DNA, in its vast majority retrotranscribed via an RNA intermediate, comprises nowadays approximately 10% of the human genome. More than 500.00 separate integration events into germ line genomes must have occured during evolution (Table 1), the results of which are now detectable as transposable ele-

Table 1: Examples of the frequency of human transposable elements (per haploid genome).

Type of element	Designation	Copy number
Retroposon	LINE1 SINE Alu family	20.000 100.000 500.000
Retrotansposon	THE1 IAP	10.000 1.000
Retrovirus	HERV-K solitary HERV-K LTRs	40-50 20.000

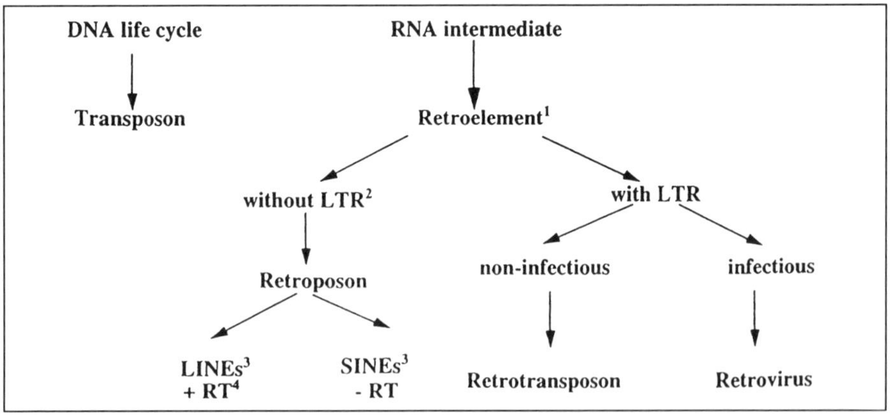

Fig. 1: Classification of transposable elements. (1) Retroelements comprise retrosequences (e.g., pseudogenes), retrotransposons and retroviruses. (2) LTR, long terminal repeats of retroviruses. (3) LINEs, long interdispersed elements; SINEs, short interdispersed elements. (4) RT, reverse transcriptase coding capacity.

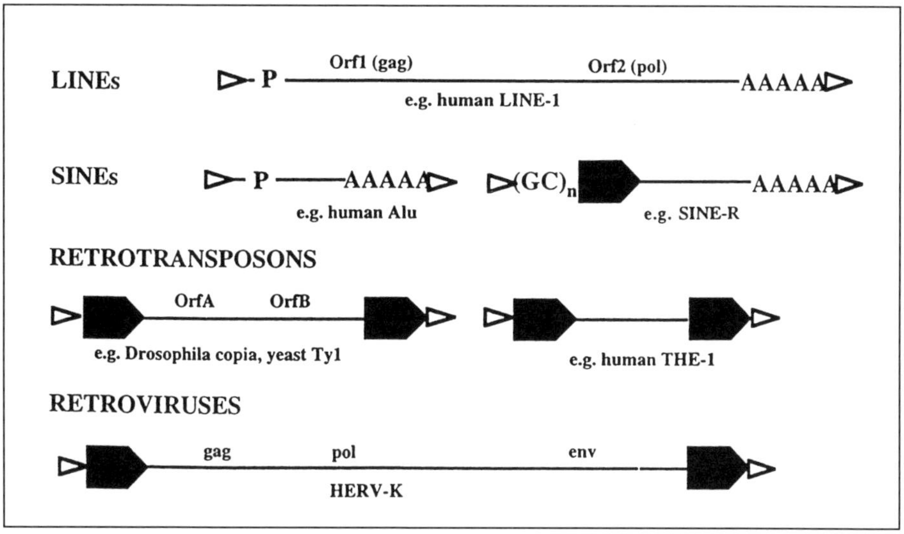

Fig. 2: Structural features of endogenous retroelements. Large arrows: long terminal repeats (LTR) of retroviruses. Small arrows: short direct repeats of cellular DNA. P: internal promoters [After 6].

ments (Fig. 1). These multicopy retroelements can easily be distinguished by their structural features (Fig. 2).

Is there anything to be learned from these very many transposition events that have ocurred in the past? Obviously, they have not been detrimental to the human species as a whole. Transposable elements are inherited like Mendelian genes. Their spread in the genome is an ongoing evolutionary process and has even been observed experimentally and clinically in animals and in humans. Thus, the lesson is that de novo insertion of DNA into chromosomes is certainly compatible with cell survival and may even confer a selective advantage. Such an event may, however, be detrimental to an individual (not to the species) if it results in malignant cell transformation (see below).

Endogenous retroviruses are biologically the most active retroelements. In some animal species they exist in an infectious form with an extracellular phase in their life cycle. By (forced) inbreeding of mice is has been observed experimentally that endogenous retroviruses can transpose («jump»), i.e. reintegrate not only in somatic, but also in germ-cell chromosomes, resulting in phenotypic mutations [2, 5]. Endogenous retrovirus sequences and particles also exist in humans [6-8]; these sequences comprise more than 1% of the human genome. Their pathophysiological potential is at present under intense scrutiny.

Retroviruses represent the only viruses that inevitably integrate almost at random into (preferably) DNAseI hypersensitive sites of the human genome. The specific molecular mechanisms ensuring chromosomal integration that are unique features of retrovirus replication will be discussed elsewhere in this volume (J.M. Coffin). Because of their consistent chromosomal integration and the safety implications associated therewith (as discussed below) it is highly unlikely that live attenuated retrovirus vaccines, for example, to protect against HIV, will be licensed, at least not in the foreseeable future. By the same token, replication-competent retroviral plasmids with inserts for DNA vaccination will not be acceptable. In contrast, replication-incompetent retroviral plasmids may well become applicable as they can be engineered to allow only one integration event upon cellular uptake in order to facilitate subsequent stablr expression of the inserted gene(s) [9].

Consequences of Foreign DNA Integration into Chromosomes

In the currently performed gene therapy clinical trials it is a common problem to infect or transfect target cells in vitro or in vivo with an efficiency that guarantees high and long-term stable expression. Therefore, one is usually aiming for chromosomal integration of the foreign replication-incompetent plasmid DNAs, regardless of whether they are of retroviral origin or not. These approaches are the more justifiable as there is a comparatively low risk in gene therapy, as discussed above.

In principle, it would also be desirable to achieve stable and high level expression of vaccine DNA which anyway encounters many imponderables once injected (Table 2). There is an inherent contradiction, however: Let's assume that a vaccine DNA is optimally expressed, i.e., stimulating both humoral and cellular immune responses. Muscle or dermal cells expressing cell surface noeantigens will inevitably come under attack from cytotoxic T lymphocytes. In other words, the more efficient (by whatever means) the DNA vaccine, the shorter the vaccination period.

It has been estimated that even under optimal in vitro experimental conditions, only about 2 out of 10^7 retroviral (!) DNA molecules are taken up by the cell [10].

Thereafter, a fraction of the foreign DNA molecules (difficult to quantitate, see below) may chromosomally integrate, which, depending on the site of integration, may

Table 2: Fate of injected DNA.

Degradation by serum nucleases
Adsorption to cells
Uptake into cells
Ligation to other DNA
Mutations
Episomal expression
Degradation by cellular nucleases
Chromosomal integration
Expression of integrated DNA
(Hyper-) methylation, resulting in inactivation
Activation or inactivation of cellular genes
Mutation, excision from chromosomal DNA, cell death

ensure expression and replication by cell division. At that stage, most molecules have acquired mutations, predominantly deletions that may endanger protein expression [11].

As indicated above, chromosomal insertion of transposable elements in all probability will rarely have detrimental consequences for the organism (Table 3). From a safety point of view, insertion of foreign DNA is meaningless as long as an inserted DNA does not become expressed, e.g., is methylated, mutated, excised or otherwise silenced, provided neighboring cellular DNA is not disturbed. Likewise, death of the cell is irrelevant for the organism. Only when chromosomal integration results in a cellular growth advantage do we encounter a risk of tumour development. Several dozens of insertional mutagenic events have been documented both for germ line and for somatic cells in animals and in humans [2, 3]. Only a few examples are listed in Tables 4 (for transposable elements) and 5 (for retroviruses). The molecular mechanisms that have been demonstrated to participate in cellular gene activation and inactivation by retroviruses are illustrated in Figure 3. Obviously, not all of these mechanisms observed after retrovirus insertion may also occur following the insertion of nonretroviral DNA that does not possess the powerful viral integrating and replicative machinery [12, 13]. For example, promoter insertion leading to down-

Table 3: Consequences of foreign DNA integration into chromosomes.

1	Insertion into biologically inert region: no expression
2	Integration into a biologically active region: stable expression
3	Stable expression, but rapid silencing (DNA methylation, mutation, excision)
4	Inactivation of vital cellular genes: cell death
5	Activation of protooncogenes (e.g., by promotor, enhancer or leader sequence insertion)
6	Creation of (tumour-specific) fusion proteins
7	Inactivation of tumour suppressor genes (e.g., by insertional mutagenesis)
8	Chromosomal instability (e.g., DNA breaks, rearrangements, changes in methylation patterns)

Table 4: Insertional mutagenesis by transposable elements..

Mobile element	Species and tissue	Locus of insertion	Effect
IAP	Murine myeloma	c-mos	Activation
IAP	Murine myelomonocytic cell line	IL-3 gene	Activation
THE1	Human germline	Calmodulin gene	?
Alu I (SINE)	Human germline	Neurofibromatosis type 1 gene	Inactivation
LINE	Human blastula	Factor VIII gene	Inactivation

Table 5: Insertional mutagenesis by retroviral integration.

Retrovirus	Tumour or cell line	Locus of insertion	Effect
Avian leukosis virus	B-cell lymphoma	c-myc	Upregulation
Mouse mammary tumour virus	Mammary adenocarcinoma	int-1 to int-4	Increase of proto-oncogene transcription
Friend murine leukemia virus	Erythroleukemia cell line	p53	Inactivation

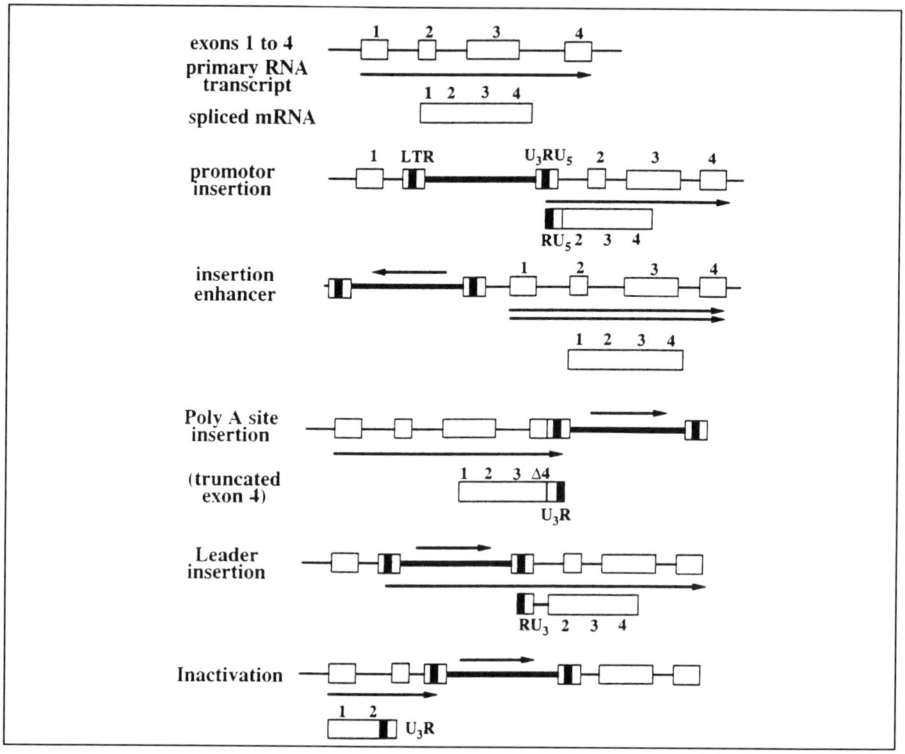

Fig. 3: Consequences of proviral (retroviral) insertion. *Promoter insertion:* down-stream transcription starting from the 3'-LTR of the inserted provirus to yeld a truncated transcript of the adjuvant cellular gene. *Enhancer insertion:* enhancer sequences of an inserted provirus activate expression of an adjuvant gene from its own promoter. *Poly A site insertion:* provirus integrates near the 3'-end of a coding region providing a poly A signal in its LTR that leads to a 3'-truncated mRNA. *Leader insertion:* read-through from provital DNA into cellular DNA resulting in a fusion product. *Inactivation:* insertional mutagenesis that gives rise to a truncated dysfunctional mRNA [12].

stream promotion of cellular genes is only conceivable if there is a strong promoter at the 3'-end of the inserting DNA.

In summary, the worrisome aspects of foreign DNA integrating into cellular DNA are those concerning the potential risk of tumour induction.

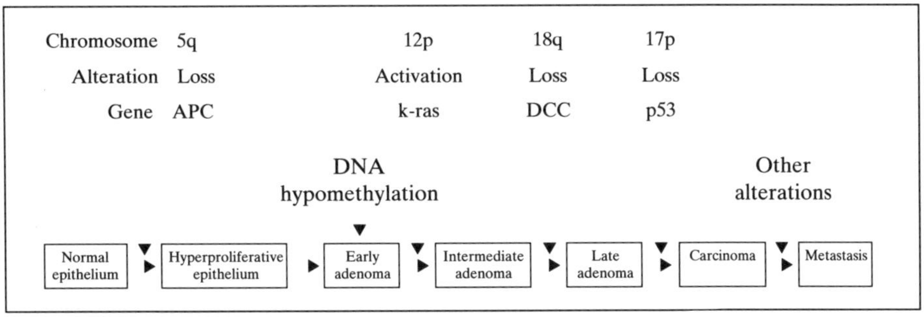

Fig. 4: A genetic model for the multistep process of colorectal tumorigenesis. APC: gene associated with adenomatous polyposis coli. DCC: gene deleted in colon cancer [Based on 14].

A Long Way to Tumour Development

Carcinogenesis is a multi/step process. In the probably best studied genetic model, colorectal tumorigenesis, Vogelstein and co-workers delineated a whole series of necessary genetic alterations for the malignant transformatin of normal epithelium into carcinoma and metastasizing cells (Fig. 4 and Ref. 14). Tumorigenesis involved, among other steps, the loss of at least three tumour suppressor gene functions and the activation of the k-ras oncogene. In other words, simple activation of one of the 100 cellular protooncogenes [reviewed in Ref. 15] or inactivation of one of the fewer tumour suppressor genes [16], of which about eight have been characterized in some detail (Table 6) is alone not yet sufficient to endow the cell with an increased growth potential.

Most protooncogenes are tissue-type specific, i.e., their insertional activation is likely to be without intracellular consequences. Notable exceptions, however, are proto-oncogenes such as p-ras and p-myc that are expressed in a wide variety of tumour cells. Protooncogene expression anyway would only have an effect if that cell had the potential to give rise to cancer. The multistep process in tumorigenesis requires cell division. Myotubes are postmytotic and other muscle cells around the site of DNA injection also possess a limited capacity for cell division. Likewise, intradermal injection would introduce vaccine DNA into differentiating cells with a limited growth potential. The notorious difficulty of immortalizing or transforming human cells in culture (except by tumour viruses) is another observation pertinent to the discussion about safety.

Activation of Protooncogenes and Inactivation of Tumour Suppressor Genes: A Numbers Game

It is very risky to try to quantify the likelihood of gene activation or inactivation caused by a chromosomal integration event of a foreign (non viral) DNA molecule [17]. The scientific literature is scant and one has to take refuge in a series of assumptions based on the results from a variety of transfection or injection studies performed mosly with retroviral DNA. Nevertheless, to establish a (certainly disputable)

Table 6: Tumour suppressor genes isolated by molecular cloning.

Designation	**Mutated in**
Rb	retinoblastoma
p53	wide range of tumours
WT-1	Wilms' tumour
NF 1	neurofibromatosis type-1
NF 2	neurofibromatosis type-2
VHL	von Hippel-Lindau syndrome
APC	adenomatous polyposis coli
DCC	deleted in colon cancer

basis for the quantification of risk factors I will try to calculate the frequency of activation of cellular protooncogenes and inactivation of tumour suppressor genes.

In a particularly relevant publication for DNA vaccination, Fung et al [1] injected the avian sarcomavirus oncogene v-src DNA (inserted into pBR322) into the wing-web muscle of susceptible chickens and subsequently observed sarcoma development in approximately 50% of the birds (TD_{50}; dose: 3×10^{11} DNA molecules). The probability of tumour formation could be calculated to be 3×10^{-12} (= 1 : (3×10^{11})) per molecule DNA (Table 7). The efficiency of cellular uptake of injected DNA in vivo may be around 10^{-9}, as under optimized conditions of in vitro transfection the probability of entrance of a retroviral DNA molecule was found to be about $2\ 10^{-7}$ [10]. If so, only 300 of the 3×10^{11} v-src DNA molecules required for tumour formation may have entered those single cells that were transformed into tumour cells. The tumour cells had integrated v-src and were monoclonal in origin. The minimum likelihood of an intracellular foreign DNA molecule entering the nucleus and becoming integrated is therefore 1 : 300 or 0.3 percent. This figure is relatively high but may nevertheless reflect the real situation: in vitro again under very optimized conditions up to 20% of intracellular DNA molecules with no special sequences promoting integration were stably inserted [18]. Calculating the target size for insertional gene activation also depends on a number of imponderables. Not all chromatin may be open for foreign DNA insertion. Retroviruses thought to be integrating at random nevertheless prefer between 100-200 yet ill-defined chromosomal regions [19]. Integration of foreign DNA into a gene may not necessarily result in (in-)activation (Fig. 3). Integration into introns may well be silent. For example, the Rb tumour suppressor gene is 190 kbp in length but has only a 2.5-kbp coding capacity [20]. However, retroviruses and expression plasmids in introns (and exons) will probably be disruptive as their polyA-signal will terminate RNA transcription prematurely.

As mentioned above, it is widely accepted that tumorigenesis is a multistep process (Fig. 4). Activation of an individual protooncogene, even if influencing the cell's phenotype and even if the cell has the intrinsic potential to become a cancer cell, is certainly not sufficient for tumour development. If one therefore calculates that at least two independent activation events are required for this unwanted side effect of DNA vaccination to materialize, the corresponding probability becomes extremely low (10^{-19}, Table 7). Comparable considerations are applicable for calculating the risk of inactivation of tumour suppressor genes by integrating foreign non viral DNA (Table 8). Again, inactivating just one allele of an appropriate tumour suppressor gene represents only one event in multistep oncogenesis. If both alleles of a relevant tumour suppressor gene need to be inactivated or if two different tumour suppressor genes must be destroyed, the probability of the two independent events occuring in the same cell is greater than 10^{-19} (Table 8).

How realistic is such speculative calculation? If the probability of the event that two DNA molecules activate two independent protooncogenes is 10^{-19}, it follows that this consequence will occur once every 10^{19} DNA molecules injected. Let us assume the DNA molecules (plasmid plus insert) are 10.000 bp in length, i.e., have a m.w. of approximately 6×10^6 d. Thus 6×10^6 gDNA correspond to 6×10^{23} molecules. 10^{19} molecules amount to $(6 \times 10^6) : (6 \times 10^{23}) \times 10^{19}$ g $= 10^2$ g $= 10^8$ µg. If a DNA vaccine dose is 10 µg, the activation event can be speculated to occur once in 10^7 vaccinees.

Well, all numbers. If, for example, the rate of integration for intracellular nonviral DNA is even smaller than the 0.3% assumed above, the risk will likewise be even smaller. If the vaccine dose is higher than the anticipated 10 µg or if the cell is already semimalignant, the risks will obviously be higher. Food for discussions.

In summary, the likelihood of tumour development in vaccinees receiving DNA free of viral or cellular oncogenes is exceedingly small and hardly calculable. Nevertheless, as we are just beginning to accumulate knowledge in the field of nucleic acid vaccination, we have to keep in mind that there is at least a theroretical risk. In the next few years we will undoubtedly be able to evaluate the results from many DNA vaccination studies in animals, and many animal species are more susceptible to cancer development than humans. Furthermore, specific research targeted at

Table 7: Activation of protooncogenes by intracellular foreign DNA.

	Probability of event
Probability of tumour formation by direct injection of viral oncogene into a sensitive host per molecule DNA[a]	3×10^{-12}
Efficiency of cellular uptake per injected DNA molecule	10^{-9}
Probability of intracellular DNA molecules to integrate chromosomally:[b] $(3 \times 10^{-12}): 10^{-9}$	3×10^{-3}
Probability of intracellular DNA molecule to integrate to integrate into a protooncogene[b]	10^{-5}
Probability of gene activation by integration	10^{-2}
Probability of a single intracellular DNA molecule to activate a single protooncogene: $3 \times 10^{-3} \times 10^{-5} \times 10^{-2}$	3×10^{-10}
At least two independent activation events required per cell $(3 \times 10^{-10})^2$	10^{-19}

a Injection of v-src plasmid DNA into the wing web muscle of young chickens:[1] measure for successful integration events.

b Calculated for intracellular DNA molecules with no special sequences promoting integration.

c Target size for insertional gene activation: 3×10^4 bp (size of a typical gene) in the 3×10^9 bp the haploid human genome.

Table 8: Inactivation of tumour suppressor genes by intracellular foreign DNA.

	Probability of event
Probability of intracellular DNA molecules to integrate chromosomally[a]	3×10^{-3}
Probability of intracellular DNA molecules to integrate into a tumour suppressor gene[b]	10^{-5}
Probability of gene inactivation by integration	10^{-2}
Probability of a single intracellular DNA molecule to inactive a single tumour suppressor gene: 3×10^{-3} to $10^{-5} \times 10^{-2}$	3×10^{-10}
At least two independent inactivation events required per cell: [c] $(3 \times 10^{-10})^2$	10^{-19}

[a] Calculated for intracellular DNA molecules with no special sequences promoting integration.

[b] Target size for insertional gene activation: 3×10^4 bp (size of a typical gene) in the 3×10^9 bp of the haploid human genome.

[c] Either to eliminate both tumour suppressor gene alleles or two different suppressor genes.

corresponding safety issues is required. One could, among many other conceivable projects, employ tumour suppressor gene knockout mice and evaluate chromosomal integration, protooncogene activation or tumour formation following the injection of murine retroviral oncogene and nonretroviral plasmid DNA. In the meantime, risk assessment of DNA vaccines should follow the extensively discussed guidelines develop for the evaluation of live virus vaccines.

For the Future: Targeting and Tethering Site-Specific Integration

Targeting foreign DNA integration or tethering it to specific DNA sequences for subsequent integration would allow a reduction in the frequency of illegitimate recombination, increase homologous recombination and add a safety margin to those approaches for which chromosomal integration is desired, for example, in ex vivo gene therapy. Site-specific DNA integration of retrotransposons such as the Ty elements of yeast cells may suggest just one strategy to avoid insertional mutagenesis. Retrotransposons are transcribed, reverse transcribed and chromosomally integrated within a single cell in a highly ordered process involving reverse transcriptase and integrase [6]. As yeast cells are haploid, retrotransposition must be precisely controlled to avoid lethal disruption of cellular genes. Thus, Ty elements display extreme biases in target site selection. Only a few weeks ago, it was shown in vitro that, for example, the Ty3 integration complex is tethered to the target DNA by the TFIIIB and TFIIIC transcription factors of Pol III polymerase within the five base pairs at the start site of transcription of polymerase III-transcribed genes [21-22]. Tethering involves protein-protein binding between at least the TFIIIB/TFIIIC factors and Ty3 integrase. It remains to be elucidated how plasmids (for example, with early genes coding for tethering proteins) or, more likely, co-plasmids can be constructed to code for those proteins that, after binding to the expression plasmid DNA, can utilize cellular factor for site-specific integration.

In a related development, Stephen Goff's group recently identified a human protein (Ini 1 for integrase interactor protein 1) that binds to the HIV-1 integrase protein (IN), which is responsible for the initial covalent binding of viral to target cell DNA [23]. The gene, designated INI, shares sequence homology with the yeast SNF5 transcriptional activator required for high-level gene expression. The Ini 1 protein also binds to cellular DNA, it stimulates the joining activity of HIV-1 IN and may target incoming viral DNA to DNAse I hypersensitive sites of active genes. Thus, tethering of the HIV integration machinery to host DNA via an IN-Ini 1 protein interaction may promote integration into preferred integration sites, already described for lymphomas with a common site of HIV integration [24]. Again, it would probably require expression of HIV-1 IN from a co-plasmid to target HIV DNA after binding to the IN-Ini 1 protein complex to preferred integration sites.

Bushman [25] already exploited tethering by fusing HIV-1 integrase DNA to the sequence-specific DNA binding domain of phage lambda repressor. The fusion protein IN-λR was successfully tested in vitro for integration site selection, supporting the idea that tethering can be sufficient to direct selective integration.

Finally, one may be able to exploit the advantages of legitimate recombination by constructing plasmids with terminal DNA sequences homologous to the preferred cellular integration sites. The problem is how to direct the plasmid to the desired integration site. It is probably hopeless to aim for single-copy cellular sequences as the plasmid DNA is unlikely to «find» them in the abundance of the 3 x 10^9 bp of the haploid human genome (unless one suceeds in finding a way to employ adenovirus-associated virus (AAV) DNA to chaperone the plasmid to the predominant chromosomal integration site AAV on chromosome 19). It is thus necessary to target repetitive sequences. Besides human retrotransposon DNA [for review, see Ref. 6], one promising family of genetic elements may be the approximately 50 copies of HERV-K endogenous retrovirus sequences and the 20,000 copies of solitary HERV-K long terminal repeats per haploid human genome [6-8]. HERV-K is expressed in many cells (unpublished observation). Plasmids containing marker genes plus HERV-K LTRs will soon be tested in this laboratory for efficiency of expression and site-directed chromosomal integration.

Acknowledgements

I would like to thank my colleagues Drs. Johannes Löwer, Klaus Cichutek, Phil Minor, Stephen Norley and Joachim Denner for vivid discussions (especially of the numbers game delineated in Tables 7 and 8) as well as Barbara Brandi and J. Denner for editorial assistance.

References

1 Fung Y-KT, Crittenden LB, Fadly AM, Kung HJ: Tumour induction by direct injection of cloned v-*src* DNA into chickens. Proc Natl Acad Sci USA 1983;80:353-357.

2 Gray DA: Insertional mutagenesis: Neoplasia arising from retroviral integration. Cancer Invest 1991;9:295-304.

3 Amariglio N, Rechavi G: Insertional mutagenesis by transposable elements in the mammalian genome. Environ Mol Mutagen 1993;21:212-218.

4 Temin MH: Reverse transcription in the eukaryotic genome: retroviruses, pararetroviruses, retrotransposons, and retrotranscripts. Mol Biol Evol 1985;2:455-468.

5 Stoye JP, Fenner S, Greenoak GE, Moran C, Coffin JM: Role of endogenous retroviruses as mutagens: The hairless mutation of mice. Cell 1988;54:383-391.

6 Wilkinson DA, Mager DL, Leong J-AC: Endogenous human retroviruses. in Levy JA (ed.): The retroviridae, New York Plenum Press, 1994, pp 465-535.

7 Löwer R, Boller K, Hasenmaier B, Korbmacher C, Müller-Lantzsch N, Löwer J, Kurth R: Identification of human endogenous retroviruses with complex mRNA expression and particle formation. Proc Natl Acad Sci USA 1993;90:4480-4484.

8 Löwer R, Tönjes RR, Korbmacher C, Kurth R, Löwer J: Identification of a rev-related protein by analysis of spliced transcripts of the human endogenous retroviruses HDTV/HERV-K. J Virol 1995;69:141-149.

9 Vogel M, Cichutek K, Norley S, Kurth R: Self-limiting infection by *int/nef*-double mutants of simian immunodeficiency virus. Virology 1993;193:115-123.

10 Kawai S, Nishizawa M: New procedure for DNA transfection with polycation and dimethylsulfoxide. Mol Cell Biol 1983;6:1787-1795.

11 Wake CT, Gudewic T, Porter T, White A, Wilson JH: How damaged is the biologically active subpopulation of transfected DNA? Mol Cell Biol 1984;4:387-398.

12 Coffin JM: Molecular mechanisms of nucleic acid integration. J Med Virol 1990;31:43-49.

13 Coffin JM: Retroviral DNA integration, in Brown F, Esber EE, Williams M (eds): Continuous Cell Lines – An International Workshop on Current Issues. Dev Biol Stand, Basel, Karger, 1992, vol 76, pp 141-151.

14 Fearon ER, Vogelstein B: A genetic model for colorectal tumorigenesis. Cell 1990;61:759-767.

15 Fan H: Retroviruses and their role in cancer, in Levy JA (ed): The Retroviridae. New York, Plenum Press, 1994, pp 313-362.

16 Hinds PW, Weinberg RA: Tumor suppressor genes. Curr Opin Genet Dev 1994;4:135-141.

17 Temin HM: Overview of biological effects of addition of DNA molecules to cells. J Med Virol 1990;31:13-17.

18 Dornburg R, Temin HM: Presence of retrovirus encapsidation sequence in nonretroviral RNA increases the efficiency of formation of cDNA genes. J Virol 1990;64:886-889.

19 Shih CC, Stoye JP, Coffin JM: Highly preferred targets for retrovirus integration. Cell 1988;53: 531-537.

20 Lee WH, Boolstein R, Hong F, Young LJ, Shew JY, Lee E: Human retinoblastoma susceptibility gene: Cloning, identification, and sequence. Science 1987;235:1394-1399.

21 Kirchner J, Connolly CM, Sandmeyer SB: Requirement of RNA polymerase III transcription factors for in vitro position-specific integration of a retroviruslike element. Science 1995;267: 1488-1491.

22 Bushman F: Targeting retroviral integration. Science 1995;267:1443-1444.

23 Kalpana GV, Marmon S, Wang W, Crabtree GR, Goff SP: Binding and stimulation of HIV-I integrase by a human homolog of yeast transcription factor SNFS. Science 1994;266:2002-2006.

24 Shiramizu B, Herndier BG, McGrath MS: Identification of a common clonal human immunodeficiency virus integration site in human immunodeficiency virus-associated lymphomas. Cancer Res 1994;54:2069-2072.

25 Bushman FD: Tethering human immunodeficiency virus 1 integrase to a DNA site directs integration to nearly sequences. Proc Natl Acad Sci USA 1994;91:9233-9237.

Dr R. Kurth, Paul-Ehrlich-Institut, Paul-Ehrlich-Str. 51-59, D-63225 Langen, Germany.

Brown F, Griffiths E, Horaud F, Petricciani JC (eds): Safety of Biological Products Prepared from Mammalian Cell Culture.
Dev Biol Stand. Basel, Karger, 1998, vol 93, pp 57-60

DNA Issues

P. Meulien

Department of Research, Pasteur Mérieux Connaught, Marcy l'Etoile, France

Summary

For many years the debate surrounding the potential danger of contaminating cellular DNA in biological products has focussed on the possibility that this DNA could give rise to tumours in the recipient host. Current knowledge of the nature and the fate of this DNA after in vivo administration allows us to conclude that this material represents a minimal risk to the patient.

Introduction

For several decades the potential safety issues arising from the presence of residual cellular DNA in biological products derived from mammalian cell cultures has been much debated [1]. Rather than re-opening the discussion from the beginning, this short presentation reviews our current understanding of the theoretical considerations related to the fate of injected DNA fragments and the accompanying risk factors for the recipient host.

For clarification, it should be stated from the outset that this discussion concerns contaminating DNA originating from cells which have the ability to grow indefinitely (resulting in the term «continuous cell lines»). Since it is known that the inherent ability of these cells to grow continuously is linked to mutations or alterations in genes regulating cell proliferation and cell development, it was throught that activated oncogenes or mutated tumour suppressor genes could be present in contaminating DNA contained in biological products derived from the type of cell substrate.

The potential danger was thus attributed to contaminating DNA fragments present in the injected product being taken up by host cells and integrated into the host chromosome. The altered chromosome could then abnormally express genes involved in tumorigenesis and lead to malignancies. This possibility was exemplified when Fung et al [2] demonstrated that tumours could be generated in chickens by directly injecting DNA encoding the oncoprotein src at an efficiency of one tumour per 3.10^{11} DNA molecules. In this case it should be noted that the 3.10^{11} molecules of DNA injected were identical and thus completely different from the heterogeneous population of DNA fragments found in products. Althrough theoretical arguments will always exist concerning the risk factors related to contaminating DNA, state of the art knowledge should now be taken into account when policies are made regarding the amounts of DNA which are permitted in biological products, especially relative to:

(i) the nature of the residual DNA in the product;

(ii) the efficiency at which this DNA is taken up by cells in vivo;

(iii) the efficiency of DNA recombination events in vivo;

(iv) the complex nature of the tumorigenic process.

Each of these points is discussed briefly to formulate a conclusion of the current thinking on the subject.

The Nature of Residual DNA in Products Derived from Continuous Cell Cultures

The mammalian cell genome contains approximately 3.10^9 base pairs of DNA per haploid genome of which only about 10% encode the 50-100,000 functional genes believed to exist. The 100pg of DNA which represent the maximum per dose quantity currently permitted in biological products would represent about 10^8 DNA fragments given that the average fargment size of the DNA in this type of product is less than 1Kb. This quantity of DNA would be equivalent to the amount contained in 10-15 genomes and by definition would be heterogeneous in nature. Since the majority of mammalian genes contain introns, the average gene size has been estimated to be about 30Kb and therefore the chance of a contaminating DNA fragment encoding a complete functional gene is extremely low. The main impact that this DNA would have on a transfected cell would thus be linked to its capacity to perturb some aspect of gene expression on the host DNA (see below).

What is the Efficiency with which this DNA Could be Transfected into Host Cells in Vivo?

Because of the recent interest in in vivo gene transfer experiments in the fields of both gene therapy and DNA vaccination, much work has now been done to assess the frequency of in vivo DNA transfer. It is thus well established [3] that in vivo DNA transfer efficiency is about one molecule in 10^9. This frequency needs to be accumulated with the frequency of recombination, allowing the integration of foreign DNA fragments into the host chromosome. There are two possibilities:

(i) illegitimate recombination which would occur at a frequency of one in 10^4 molecules;

(ii) homologous recombination for which the frequency is roughly three orders of magnitude less (one in 10^7 molecules) [4].

The Consequences of DNA Integration

When DNA integrates into the chromosome, several outcomes are possible and these include: (i) insertion into a biologically inert region of the chromosome (the most likely as this represents 90% of the genome) without any particular consequence; (ii) insertion into a coding region where genes could be activated, inactivated, up-regulated or down-regulated.

A potential risk of tumour development would only occur if the target gene (among the 50,000 or 100,000 present) was associated in some way with the tumorigenic process. This process is in itself highly complex as a cascade of events is necessary before cells can become malignant. Temin [5] suggested that the target for integration would be about 30 Kb (average gene size) and estimated that about 100 of the 50-100,000 genes present in the genome (one in 10^3) could be associated with oncogenesis. Given the results already cited concerning the tumorigenicity of src DNA (being one in 3.10^{11} molecules) then the risk of a tumour being induced by incoming contaminating DNA integrating near proto-oncogene and leading to the activation of this oncogene would be in excess of one in 10^{16} (10^{11} x 10^3 x 10^2) and would probably be closer to the range of one in 10^{18} to 10^{21} (per molecule of DNA introduced) as calculated in Table 1. This worst case scenario gives some idea of the extremely low risk that contaminating DNA would have in inducing any detrimental effect in the recipient. Indeed, by any calculation the risk would be well below the natural mutation rate estimated to be one in 5 x 10^5 genes.

Table 1: Overall risk of contaminating DNA inducing tumours.

Frequency of uptake of DNA in vivo	10^{-9}
Frequency of recombination	10^{-4}-10^{-7}
Frequency of integration into coding region	10^{-3}
Frequency of integration near or in a proto-oncogene	10^{-2}
Accumulated risk	10^{-18}-10^{-21}

Conclusion

In this resumé I have tried to summarize our current knowledge of the risk factors associated with contaminating DNA in biological products. Recent clarification on the fate of DNA molecules injected in vivo has been made essentially by the work on gene delivery in the gene therapy and DNA vaccine fields. Indeed, doses of about 100 µg of plasmid DNA in mice and up to several milligrams in non-human primates have been administered in attempts to induce some immune responses. Biodistribution studies have also been performed (unpublished data) and it is clear that most of the injected DNA is degraded very rapidity before transfection into host cells. Coupled to these observations is the fact that low efficiencies of both transfection and DNA integration into the host genome in vivo further lower the risk of danger for the recipient.

In conclusion therefore it would seem that the monitoring of contaminating residual DNA in biological products is a useful exercise with respect to the characterization of the purity of the product and an assessment of the reproductibility of a given manufacturing process. The small amounts of DNA present in these products presents no real danger to the recipient but should be quantified as accurately as possible.

Acknowledgements

I would like to thank Luc Aujame for critical reading of the manuscript.

References

1 Petricciani JC: Cells, science and health, in Hayflick L, Hennessen W (eds): Continuous Cell Lines as Substrates for Biologicals. Dev Biol Stand. Basel, Karger, 1988, vol 70, pp 3-10.

2 Fung YK, Crittenden LB, Fadly AM, Kung HJ: Tumor incidence by direct injection of cloned v-src DNA into chickens. Proc Natl Acad Sci USA 1983:80:353-357.

3 Kurth R: Risk potential of the chromosomal insertion of foreign DNA : in DNA vaccines. Annals of the New York Academy of Sciences, 1995, vol 772.

4 Coffin JM: Molecular mechanisms of nucleic acid integration. J Med Virol 1990;31:43-49.

5 Temin HM: Overview of biological effects of addition of DNA molecules to cells. J Med Virol 1990;31:123-127.

Dr. P. Meulien, Research and Development, Pasteur Merieux Connaught Canada,
1755 Steeles Ave. West, North York, Ont., Canada M2R 3T4.

Brown F, Griffiths E, Horaud F, Petricciani JC (eds): Safety of Biological Products Prepared from Mammalian Cell Culture.
Dev Biol Stand. Basel, Karger, 1998, vol 93, pp 61-64

Residual Host Cell Protein from Continuous Cell Lines

Effect on the Safety of Protein Pharmaceuticals

J.H. Lupker

Sanofi Recherche, Centre de Labège, Labège, France

Abstract: The presence of residual host cell proteins in protein pharmaceuticals obtained from continuous mammalian cell lines has been a point of concern. Clinical experience with different biopharmaceuticals shows that residual protein in these highly purified products does not represent a danger to the health of the patients.

In the past, considerable concern has been expressed regarding the deleterious effects of host cell proteins originating from permanent – immortal – mammalian cell lines which are present in the final product. These concerns mainly referred to oncogene encoded proteins which were suspected to be a risk because of tumour induction after injection. During the previous WHO Study Group meeting on Biologicals held in Geneva in 1986, Lowy [1] came to the conclusion that the only potential risks of host cell protein are from growth factors secreted by the cell substrates. However, quantitative considerations led him to conclude that very high quantities of these growth factors would be needed to induce tumours in the recipient. For the highest excreting cells 50-75 ng/ml of growth factors are found in the medium, concentrations which are not considered to be capable of inducing tumours. The conclusion of the 1986 meeting was: *The apparent risk from oncogene-encoded proteins is limited to growth factors that can exert their biological effects via the cell surface. Growth factors may be secreted by cells in which biologically-derived reagents might be produced, but the risks from growth factors are limited, since their growth-promoting effects are usually transient and reversible* [2].

In conclusion, the presence of growth factors in low concentrations was not considered a serious risk. The consequence of this conclusion is that host cell proteins derived from continuous cell lines can be considered as ordinary protein contamination or, perhaps better, residual proteins like those present in proteins produced in micro-organisms. The same technologies can be applied to follow the purification process and to monitor impurities in the final product.

Several new protein-based pharmaceuticals produced in continuous cell lines have come to the market since 1986, and even more are under development [3]. Consequently, considerable clinical experience exists and the evaluation of the safety of these products is possible. We can see whether any deleterious effect due to contaminating proteins has been reported in the open literature that could invalidate the optimistic conclusions of the 1986 meeting.

To monitor both the removal of protein impurities during the process and their presence in the final product, special immunological methods have been developed for *E. coli*-derived proteins [4]. Later, these methods were adapted to proteins produced in mammalian cells [5]. These tests in general use a control cell line which carries an expression vector similar to that used for the production of the desired protein, with the exception of the cDNA coding for the protein of interest. The cell line is grown under conditions comparable to those used during a production run. A similar purification process is then performed which yields different pools of contaminating protein. These are used for the production of antisera specific for host cell proteins.

In this way it is possible to obtain very sensitive immuno-assays for the detection and quantification of protein impurities in the final product [6]. It is generally considered that these methods are superior to in vitro and in vivo testing for the detection of possibly mutagenic and/or dangerous contaminants of protein pharmaceuticals [7]. One point, however, should be made: the immunological tests used not only detect residual host cell proteins, but also residual proteins which are added in the production process, such as serum proteins or proteins added to the culture medium (e.g. albumin, insulin or transferrin). The immunological assays, together with electrophoretic analysis and rigorous process validation clearance studies, are considered to be the basis for the quality assessment of protein pharmaceuticals at least as far as protein impurities are concerned [6]. The careful process design combined with rigorous process validation and GMP production will assure the consistency and safety of protein pharmaceuticals produced in mammalian cells.

At the end of 1996, different host cell lines have been used to produce protein pharmaceuticals, notably Chinese Hamster Ovary cells, Baby Hamster kidney cells and murine C127 cells. Some of the products made with these cells have been on the market for several years. For these proteins, reports are published in the literature evaluating their possible adverse clinical effects. However, care should be taken in the evaluation of this literature. Since most of the products are extremely active biological substances, it is difficult to evaluate which part of the adverse effects is due to the contaminants and which part is due to the active substance itself. Moreover, some proteins are produced both in micro-organisms and in mammalian cells. Reports in the literature on possible adverse effects quite often do not discriminate between the two.

Human growth hormone, or Somatotropin, is one of the drugs produced in both *E. coli* and mammalian cells. Treatment with the hormone produced in mammalian cells (C129) did not result in antibody production to host cell antigens [8, 9]. The question has been raised whether the treatment with human growth hormone increases the risk of malignancies [10]. In subsequent epidemiological studies, no indication for this was found [11]. In this case there is no distinction in the literature between patients treated with bacterial or mammalian cell derived human growth hormone.

Adverse effects of treatment with recombinant human erythropoietin (EPO) have been described. Only in very rare cases has an anaphylactic reaction been reported [12]. This seems to be caused by the development of antibodies to the recombinant human erythropoietin but not to host cell-derived protein impurities.

Two different recombinant Factor VIII products are on the market. A number of adverse immunological reactions have been described [13, 14] but they do not seem to be related to the recombinant form of this protein, because the same kind of effects are found in plasma-derived Factor VIII preparations. No immunological reaction to host cell proteins has been documented, neither in long term studies [13, 14] nor in a reported case of anaphylaxis [15].

Another product for which much clinical experience exists is a recombinant human Tissue Plasminogen Activator (TPA). In this example also there is no indication that antibodies are formed against potential residual host-cell protein [16].

Other products made with continuous mammalian cell lines are on the market, or are currently in clinical trials. For those products, adverse effects are reported but they are in general related to the biological activity of the protein. No antibodies have been detected in these cases, for example to the recombinant forms of human Granulocyte Colony Stimulating Factor [17], Follicle Stimulating hormone [18] or Thyrotropin [19], nor has mention been made of antibodies to residual host-cell proteins.

Careful production process design, development of appropriate analytical tools, rigorous process validation, and application of Good Manufacturing Practice, which is current practice in the biopharmaceutical industry, have resulted in pure and consistent products. Clinical experience shows us that the concern that residual host-cell proteins in biopharmaceuticals produced with continuous mammalian cells are a danger for the patients is no longer justified.

References

1 Lowy DR: Potential oncogenic hazards posed by oncogene encoded proteins, in Petricciani JC, Hennessen W (eds): Cells, Products, Safety. Dev Biol Stand. Basel, Karger, 1987, vol 68, pp 63-67.

2 WHO meeting on cells for the production of biologicals, in Petricciani, Hennessen W (eds): Cells, Products, Safety. Dev Biol Stand. Basel, Karger, 1987, vol 68, pp 69-72.

3 Lubiniecki AS, Lupker JH: Purified protein products of rDNA technology expressed in animal cell culture. Biologicals 1994;22:161-169.

4 Anicetti VR, Fehskens EF, Reed BR, Chen AB, Moore P, Geier MD, Jones AJS: Immunoassay for the detection of E.coli proteins in recombinant DNA derived human growth hormone. Journal of Immunol Methods 1986;91:213-224.

5 Pauly JU, Siebold B, Schulz R, List W, Lüben G, Seiler FR: Development of an ELISA for the detection and determination of contaminating proteins in recombinant DNA derived human erythropoietin. Behring Inst Mitt. 1990;86:192-207.

6 Eaton LC: Host cell contaminant protein assay development for recombinant biopharmaceuticals. J Chrom A 1995;705:105-114.

7 Teelmann K, Hohbach C, Lehmann H, The International Working Group: Arch Toxicol 1986; 59:195-200.

8 Zeisel HJ, Lutz A, v Petrykowski W: Immunogenicity of a mammalian cell-derived recombinant human growth hormone preparation during long-term treatment. Horm Res 1992;37 (suppl2): 47-55.

9 Frasier SD, Rudlin CR, Zeisel HJ, Liu HH, Long PC, Senior B, Finegold DN, Bercu BB, Marks JF, Redmond P: Effect of somatotropin of mammalian cell origin in growth hormone deficiency. Am J of Diseases of Children 1992;146:582-587.

10 Ritzen EM: Does growth hormone increase the risk of malignancies? Horm Res 1993;39:99-101.

11 Blethen SL: Complications of growth hormone therapy in children. Current Opinion in Pediatrics 1995;7:466-471.

12 Garcia JE, Senent C, Pascual C, Fernandez G, Perez-Carral C, Diaz-Tejeiro R, Gomez E, Sierra T: Anaphylactic reaction to recombinant human erythropoietin. Nephron 1993;65:636-637.

13 Lusher JM: Summary of clinical experience with recombinant factor VIII products – Kogenate. Ann Hematol 1994;68:S3-S6.

14 White II GC: Summary of clinical experience with recombinant factor VIII products – Recombinate. Ann Hematol 1994;68:S7-S8.

15 Shopnick RI, Kazemi M, Brettler DB, Buckwalter C, Yang L, Bray G, Gomperts ED: Anaphylaxis after treatment with recombinant factor VIII. Transfusion 1996;36:358-361.

16 Reed BR, Chen AB, Tanswell P, Prince WS, Wert Jr. RM, Glaesle-Schwarz L, Grossbard EB. Low incidence of antibodies to recombinant human tissue-type plasminogen activator in treated patients. Thromb Haemost 1990;64:276-280.

17 Sasaki O, Yokoyama A, Uemura S, Fujino S, Inoue Y, Kohno N, Hiwada K: Drug eruption caused by recombinant human G-CSF. Internal Medicine 1994;33:641-643.

18 Fisch B, Avrech OM, Pinkas H, Neri A, Rufas O, Ovadia J, Loumaye E: Superovulation before IVF by recombinant versus urinary human FSH (combined with a long GnRH analog protocol): a comparative study. J of Assisted Reproduction and Genetics 1995;12:26-31.

19 Meier CA, Braverman, LE, Ebner SA, Veronikis I, Daniels GH, Ross DS, Deraska DJ, Davies TF, Valentine M, DeGroot LJ, Curran P, McEllin K, Reynolds J. Robbins J, Weintraub BD: Diagnostic use of recombinant human thyrotropin in patients with thyroid carcinoma (Phase I/II study). J Clin Endocrinal Metab 1994;78:188-196.

Dr. J.H. Lupker, Sanofi Recherche, Centre de Labège, B.P. 137, F-31676 Labège Cedex, France.

Brown F, Griffiths E, Horaud F, Petricciani JC (eds): Safety of Biological Products Prepared from Mammalian Cell Culture.
Dev Biol Stand. Basel, Karger, 1998, vol 93, pp 65-72

An Overview of the BSE Epidemic in the UK

R. Bradley

Central Veterinary Laboratory, New Haw, Addlestone, UK

Key word: BSE, spongiform encephalopathy, prion disease.

Abstract: The BSE epidemic in the UK is declining rapidly in response to control measures and is en route for eradication by the year 2001. BSE infectivity is restricted in clinically affected natural cases of BSE to central nervous tissues. BSE agent has been isolated from the intestine of experimentally orally-infected cattle from six to 18 months post-challenge. A small maternal risk factor for BSE has been identified but this will not stop the eradication of BSE. To protect the integrity of biological products from the risks associated with the unknown hazard of BSE for man it is recommended that only safe sources of animals and tissues are used. Inactivation procedures employed during processing will provide further guarantees.

Introduction

Bovine Spongiform Encephalopathy (BSE) is a new disease of domestic cattle, especially dairy cattle, first discovered in Great Britain in 1985 [1]. BSE is a member of the group of diseases known as the sub-acute, transmissible spongiform encephalopathies (TSE) or prion diseases. In this regard BSE shares some features with scrapie of sheep and goats, which has been known for over 250 years and Creutzfeldt-Jakob disease (CJD) of man which has been known for over 70 years. However, despite much epidemiological investigation, no epidemiological link has ever been made between TSE of animals and CJD of man, or vice versa, either as a result of natural or iatrogenic transmission. Indeed, other than in scrapie there is little evidence for natural animal to animal, or for human to human spread of these diseases. However, there has been a small number of tragic, but preventable iatrogenic transmissions between humans (CJD) and one in sheep (scrapie). In man, with few exceptions, all have been due to the use of biological products contaminated with the CJD agent. It is the need for developing procedures for preventing iatrogenic disease via medicinal products that is the focus of attention in this conference.

The recently reported occurrence over a few months in 1995 and 1996 of ten cases of a new variant form of sporadic CJD (NV-CJD) in young people, including teenagers [2], has resulted in renewed concerns for human health from the hazard of BSE. Historically, teenagers and young people have been, extremely rarely affected by conventional, sporadic CJD. In the absence of establishing familial or iatrogenic CJD as the cause of NV-CJD, the most likely explanation is that it is caused by a strain of CJD agent not previously encountered or reported. The public has largely been concerned about dietary exposure. Some authorities, including the UK Spongiform Encephalopathy Advisory Committee (SEAC), have suggested that the most likely origin is exposure before 1989 to infected central nervous tissue from cattle with BSE. It was in September 1989 that a ban on Specified Bovine Offals (SBO), which include brain and spinal cord, was introduced in the UK so they could no longer enter the human food chain.

Currently there is no direct evidence of a link between NV-CJD and BSE. However, it remains a possibility that there is such a link but this is unlikely to be proved or disproved, at least until the outcome of further epidemiological studies on the existing and any new cases of NV-CJD (four further cases in the UK and one in France have been confirmed and reported to 30 September 1996). This could take several more years as the mean incubation period is unknown and could be a decade or more. Other indications that there may be, or there is not, a causal connection could result from the outcome of comparative strain-typing of isolates from patients with NV-CJD and from cattle with BSE in conventional mice and the results of transmission studies in transgenic mice carrying various forms of the human *PrP* gene. It is possible that no study will be conclusive in the short term and therefore there will probably be a degree of uncertainty for many years. The way forward therefore is to ensure that material collected from animals for use in biologicals or during their manufacture is free from the infectious agent by selecting tissue from safe sources. In addition it is important to ensure that processing procedures are consistently as effective as is possible at inactivating TSE infectivity. The former is the essential component of the belt and braces policy in regard to TSE infection control.

The purpose of this paper is to provide the basic evidence upon which can be based the necessary assurances for the safety of the source bovine materials used in, or during, the manufacture of medicinal products.

The Geographical Distribution of BSE

The UK is the only country with a high incidence of disease affecting about one per cent of adult cattle at the peak of the epidemic in 1992. This is declining rapidly and is now below 0.33 per cent [3]. Most countries in the world have not reported BSE though active surveillance in some of these countries could be questioned. Some countries in Australasia and South America can provide excellent reassurances that neither scrapie nor BSE exist there. A small number of countries (Canada, Denmark, Falkland Islands, Germany, Italy and Sultanate of Oman) have had fewer than a total of 20 cases and every one of these had been exported from the UK at an age when it was most likely to have been already infected. These cases can be regarded as an extension of the UK epidemic. France and Switzerland have had cases only in native-born cattle whereas Portugal and Ireland have had confirmed cases in cattle both exported from the UK and in native-born cattle. Switzerland has the next highest number of cases (224 to 1 August 1996) to Great Britain (163,360 to 13 September 1996), but along with all the other countries with native-born cases is regarded as having a low incidence of disease.

Strain of Agent

In the UK and in Switzerland the consistency of the clinical signs and pathology in cattle and results of strain typing in mice suggest the same major strain of agent is in each source and is different from known strains of scrapie [4]. Isolates from cattle from other countries have not been strain-typed.

New TSE in Other Species

Since the advent of BSE in 1986, and in one case just before the recognition of BSE, previously unrecorded forms of TSE have been reported in seven species of captive wild *Bovidae* in the UK (18 cases in all), in domestic cats (75 cases in the UK, one in a native-born cat in Norway) and four species of captive wild *Felidae* (10 cases in all). Two of four cheetahs developed TSE after export from Great Britain, one in Ireland, the other in Australia. The characteristics of the strain of agent from three domestic cats, a nyala and a kudu were very similar to those of the BSE strain isolated from cattle, suggesting a common origin [4].

The Origin of the BSE Epidemic in the UK

Extensive epidemiological studies have revealed that the origin of the BSE epidemic in the UK was a scrapie-like agent consumed in concentrate feed containing meat and bone meal [5]. The scrapie-like agent could have been derived from a scrapie-like agent from sheep, or a cattle-adapted form of a scrapie-like agent [6]. In any event, commercially motivated changes in processing of ruminant waste by rendering, especially in the early 1980s, coincided with the putative period of first effective exposure (1981/82) which led to clinical disease in 1985/86 [6]. **Note:** epidemiological studies revealed a few cases of BSE retrospectively following the first histopathological confirmation in 1986. A major factor in the development of the epidemic of BSE, evident as an exponential rise in clinical cases from mid-1989, was the recycling of infection from cattle to cattle via rendering and feed from about 1984. Recycling also aided the selection of cattle-adapted agent irrespective of its source [7]. If the original origin was sheep then a major obstacle, the species barrier, was overcome. Association of the occurrence of BSE with changes in rendering was further supported by studies of the effectiveness of different processing procedures used in the Member States of the European Union (EU) in inactivating BSE agent infectivity [8]. Some processes were ineffective and as a result these were banned from rendering ruminant protein in 1994 in all Member States.

All the «new» TSE in the species mentioned in the section above are believed to have resulted from feed, the ruminant species via meat and bone meal and the captive wild cats following consumption of infected uncooked central nervous tissue of cattle. The precise source of disease in domestic cats is unknown but only one case has developed in *Felidae* born after 1990, the date of the SBO ban applied to animal and bird feed [D Matthews, personal communication].

The Experimental Host Range of BSE

Domestic chickens and six mammalian species have been challenged by the intracerebral (i/c) and other routes of parenteral inoculation. Neither chickens nor hamsters succumbed, nor did chickens following oral challenge. However, transmission did occur in mice, cattle, sheep, goats, pigs, mink, marmosets and macaque monkeys [9]. Primates have not been challenged orally and pigs challenged orally have not developed BSE after 75 months. The other six species have succumbed following oral challenge but the incubation periods were longer than following i/c challenge despite the dose being substantially larger.

The Distribution of BSE Infectivity in Tissues of Cattle

Natural disease. Infectivity has been detected only in the brain, cervical and terminal spinal cord [10] and in the retina [H Fraser, DM Taylor personal communication]. The i/c mouse bioassay was used. No infectivity was found in any lymphoreticular or 40 other tissues that were tested using the same type of susceptible mouse.

Experimental disease. Following oral challenge of 30, four month old calves with 100g of brain from clinically-affected, confirmed cases of BSE, sequential kills were undertaken at approximately four month intervals commencing at six months of age. Three challenged cattle and one undosed control were killed at each time point and over forty tissues collected for bioassay in mice, avoiding risks of cross contamination. The results are known for inoculated tissues from cattle killed up to 18 months post-challenge (22 months of age). Only the distal ileum in all kills except the first has shown detectable infectivity [GAH Wells, personal communication]. Some of the cattle showed clinical signs of BSE and some had a pathological diagnosis of spongiform encephalopathy at between 35 and 37 months from challenge [GAH Wells, personal communication].

In an attack rate study, cattle in all four orally challenged groups have succumbed to BSE but there are survivors in the lower dose groups. The inoculum was undiluted brain material from clinically affected, confirmed cases of BSE. The doses were 3 x 100g; 1 x 100g; 1 x 10g and 1 x 1g [GAH Wells, personal communication].

A criticism of the mouse bioassay is that it may be insufficiently sensitive for detecting low levels of infectivity that may theoretically exist in non-central nervous tissues, despite the fact that mice challenged with brain tissue from affected animals from a variety of sources have consistently developed disease. To overcome this criticism a comparative bioassay is in progress using mice and cattle. Separate pools of brain, spleen and a range of lymph nodes have been prepared from five cattle with BSE and homogenates assayed by i/c inoculation. The titration reveals a titre of $10^{3.3}$ i/c ID_{50}/g in mice whereas interim results from cattle show the titre is likely to be c. > 100-fold higher [GAH Wells, personal communication]. No transmission has resulted in mice following inoculation of either spleen or lymph node and cattle remain alive at 43 months of age, which is beyond the age at which any cattle have succumbed to BSE following inoculation of diluted brain homogenate [GAH Wells, personal communication]. This suggests that the titre in spleen and lymph nodes of clinically BSE-affected cattle is likely to be at least one million times lower than it is in brain from the same animals. Infectivity in other non-lymphoreticular, non-central nervous tissues is likely to be even lower. It is important to emphasise that several of the above experiments are still in progress and that a final interpretation can only be made at their conclusion.

Cattle to Cattle Transmission

There is little evidence for cattle to cattle transmission. The within herd incidence has remained below three per cent in each six-month period throughout the epidemic, rising slightly as the epidemic developed and now declining [3]. Furthermore, around 70 per cent of farms that have experienced BSE have had four cases or less and about 35 per cent have had only once case [D Matthews, personal communication]. This supports the view that cattle to cattle transmission is rare or absent. Since there is little variation in the bovine *PrP* gene (which in other species plays a major role in determining susceptibility to, and incubation period of, diseases caused by TSE agents) it is not a major factor in the occurrence of BSE [7]. It follows therefore, that if there is little genetic variation in susceptibility, the generally low incidence of BSE in affected herds is due to a low average feedborne infection [11]. The same general principles apply also to BSE in other countries.

No infectivity has been found in semen, embryos, reproductive tissues, placenta or milk by bioassay in susceptible mice [3, 10]. Oronasal challenge of cattle with placenta from BSE-affected cattle has also produced no disease at over 80 months from challenge [GAH Wells, personal communication]. Neither has disease resulted in recipient cattle from New Zealand or offspring derived from embryos collected from BSE-affected cattle and transferred into these recipient cattle [AE Wrathall, personal communication]. However, although the oldest offspring are over four years old, the experiment will not be complete until 2001.

Epidemiological studies have indicated that even if maternal transmission does occur it will be at a low level and could not sustain the epidemic in the UK [12] and in a powerful case control study it is concluded that neither maternal nor horizontal transmission can account for the majority of BSE cases in cattle born after July 1988 [13] (see Control of BSE section). The recent evaluation by the SEAC of the interim results of a cohort study involving 300 pairs of cattle, one of offspring from BSE infected animals and the other control animals from the same herd and cohort whose dams were over six years old and did not develop BSE, shows a maternal risk factor for BSE in the study of the order of 10 per cent [JW Wilesmith, personal communication]. It did not indicate what the risk factor was i.e. whether it was an environmental or genetic factor and, if environmental, what the pathogenesis was. The final results from the study will not be available until after November 1996 when the last animal is killed. It may then be possible to analyse whether or not the maternal risk factor is constant or varies with the year of birth of the cattle in the study [JW Wilesmith, personal communication].

The Species Barrier Revisited

The species barrier is the main obstacle to experimental infection of other species [14] and probably also in natural disease. An effective dose is a function of the infectivity titre of the contaminated material, amount consumed or inoculated, the route of exposure and the species barrier. The i/c route is (in mice) about 10^5 times more efficient than the oral route [7]. The species barrier comprises two components, the agent strain and the variation in the *PrP* gene sequence of the donor and recipient species. In practice only the dose (the weight of tissue multiplied by its infectivity titre/g) can be manipulated practically to reduce any infectivity in the final product (e.g. by identifying safe sources and by using methods showing effective inactivation). The role of the species barrier is being investigated by various methods including the use of transgenic animals. We already know that the BSE agent inoculated into mice expressing mouse or human PrP only produce detectable mouse PrP [15]. So far, in incomplete studies, no spongiform encephalopathy has resulted following i/c inoculation of BSE into transgenic, PrP null mice expressing human PrP. These mice are now over 500 days old (J Collinge, personal communication) i.e. over twice the incubation period resulting from the successful transmission following challenge with CJD agent [15].

Control of BSE

Control of animal health was initiated in 1988 and the most important scientific measure was the ban on feeding ruminant protein to ruminant animals in July of that year. This was very effective, but not completely so, resulting in the occurrence of nearly 29,000 cases of BSE in cattle born after the ban to September 1996 [D Matthews, personal communication]. A large number (perhaps half) might be attributed to infected feed in the supply chain. The remainder can largely be attributed to accidental cross contamination of ruminant diets with feed or ingredients used in pig and poultry rations which could have contained meat and bone meal prepared from SBO (i.e. those like brain, retina and spinal cord that may have contained the infectious agent). A relatively small number of cases could theoretically be attributed to the maternal risk factor revealed in the cohort study.

Apart from the leakage in the ruminant feed ban another factor was the incomplete removal of all SBO, particularly brain, eyes and spinal cord from the material that was rendered to produce meat and bone meal for use in pig and poultry rations. A leakage of BSE infectivity into feed could have occurred at least until 1994 when the unsatisfactory rendering processes were banned. It seems impossible to prevent cross contamination of ruminant rations in mills and on farms unless dedicated machinery and transport is used. The low dose of infected brain (1g) which can cause BSE in cattle by the oral route makes it difficult to close the loophole. As a result, following the occurrence of NV-CJD in 10 patients announced in March this year, the SEAC recommended that mammalian meat and bone meal be banned for use in food animal diets including those of horses and pigs in the UK. Furthermore, the UK Government announced that no cattle over the age of 30 months would be permitted to enter the human food or animal feed chain. This measure replaced the deboning procedure recommended by the SEAC. A number of other measures have also been taken to protect public health, namely the proscribing of cattle heads, except the tongue, as specified bovine material and controls on the production of mechanically recovered meat from bovine vertebrae (excluding the tail). Additional animal health protection measures will include stricter criteria for rendering ruminant waste in the EU (133°C, at 3 bar for 20 minutes). This decision results from the analysis of interim data from a further study of rendering processes in the EU using a scrapie spike, instead of BSE (DM Taylor, personal communication).

A recent analysis of the epidemiological data from the UK epidemic [12] indicates that new infections of cattle from feed will be very rare from 1994, and these are more likely to result from a maternal risk factor. These will not sustain the epidemic which is predicted to die out naturally without further intervention by 2001 [12].

Conclusions

The BSE epidemic in the UK is likely to be concluded for practical purposes in 2001. Risks of new infections from feed in the UK are very low since 1994. Maternal transmission if it occurs will not sustain the epidemic. The same confidence cannot be provided for BSE in other countries where the lessons from the UK epidemic are being learned late. Nevertheless, the incidence of BSE in Europe outside the UK is much lower, so the risks are accordingly lower too, though they are not zero. From the data provided it should be possible to conduct sound risk assessments for the sourcing of bovine tissues for the preparation of biologicals. The objective should be to ensure that source animals and their tissues are devoid of TSE agent infection. Without any available tests for infection this has to be done by sound risk assessment methods. Reliance should not be placed upon inactivating procedures used during manufacture alone but these will provide additional guarantees that the end product is safe.

Acknowledgements

The author particularly thanks Dr H Fraser, Dr M Bruce, Dr DM Taylor of the Institute for Animal Health Neuropathogenesis Unit in Edinburgh, Dr D Matthews and Mr KC Taylor of the Ministry of Agriculture, Fisheries and Food at Tolworth, Mr GAH Wells, Mr JW Wilesmith and Dr AE Wrathall for access to unpublished data and other guidance. Any views expressed are the responsibility of the author.

References

1 Wells GAH, Scott AC, Johnson CT, Gunning RF, Hancock RD, Jeffrey M, Dawson M, Bradley R: A novel progressive spongiform encephalopathy in cattle. Vet Rec 1987;121:419-420.

2 Will RG, Ironside JW, Zeidler M, Cousens SN, Estibeiro K, Alperovitch A, Poser S, Pocchiari M, Hofman A: A new variant of Creutzfeldt-Jakob disease in the UK. The Lancet 1996;347:921-925.

3 MAFF: A Progress Report, May 1996. Ministry of Agriculture, Fisheries and Food, Tolworth UK. 1996, pp 1-51 + appendices.

4 Bruce M, Chree A, McConnell I., Foster J, Pearson G, Fraser H: Transmission of bovine spongiform encephalopathy and scrapie to mice: strain variation and the species barrier. Phil Trans Roy Soc B 1994;343:405-411.

5 Wilesmith JW, Wells GAH, Cranwell MP, Ryan JBM: Bovine spongiform encephalopathy: epidemiological studies. Vet Rec 1988;123:638-644.

6 Wilesmith JW, Ryan JBM, Atkinson MJ: Bovine spongiform encephalopathy: epidemiological studies on the origin. Vet Rec 1991;128:199-203.

7 Kimberlin RH: A scientific evaluation of research into bovine spongiform encephalopathy (BSE), in Bradley R, Marchant B (eds): Transmissible spongiform encephalopathies. Proceedings of a consultation on BSE with the Scientific Veterinary Committee of the Commission of the European Communities held in Brussels from 14-15 September 1993 CEC Brussels, 1994, pp 145-159.

8 Taylor DM, Woodgate SL, Atkinson MJ: Inactivation of the bovine spongiform encephalopathy agent by rendering procedures. Vet Rec 1995;137:605-610.

9 Lasmézas CI, Deslys J-P, Demaimay R, Adjou KT, Lamoury F, Dormont D, Robain O, Ironside J, Hauw J-J: BSE transmission to macaques. Nature 1996;381:743-744.

10 Fraser H, Foster JD: Transmission to mice, sheep and goats and bioassay of bovine tissues, in Bradley R, Marchant B (eds): Transmissible spongiform encephalopathies. Proceedings of a consultation on BSE with the Scientific Veterinary Committee of the Commission of the European Communities held in Brussels from 14-15 September 1993 CEC Brussels, 1994, pp 145-159.

11 Kimberlin RH, Wilesmith JW: Bovine spongiform encephalopathy: epidemiology, low dose exposure and risks. Annals of the New York Academy of Sciences 1994;724:210-220.

12 Anderson RM, Donnelly CA, Ferguson NM, Woolhouse MEJ, Watt CJ, Udy HJ, MaWhinney S, Dunstan SP, Southwood TRE, Wilesmith JW. Ryan JBM, Hoinville LJ, Hillerton JE, Austin AR, Wells GAH: Transmission dynamics and epidemiology of BSE in British cattle. Nature 1996;382:779-788.

13 Hoinville LJ, Wilesmith JW, Richards MS: An investigation of risk factors for cases of bovine spongiform encephalopathy born after the introduction of the 'feed ban'. Vet Rec 1995;136:312-318.

14 Kimberlin RH: An overview of bovine spongiform encephalopathy, in Brown F, Horaud F (eds): Virological Aspects of the Safety of Biological Products. Dev Biol Stand. Basel, Karger, 1991, vol 75, pp 75-82.

15 Collinge J, Palmer MS, Sidle KCL, Hill AF, Gowland I, Meads J, Asante E, Bradley R, Doey LJ, Lantos PL: Unaltered susceptibility to BSE in transgenic mice expressing human prion protein. Nature 1995;378:779-783.

Mr. R. Bradley, Central Veterinary Laboratory, New Haw, Addlestone KT15 3NB, U.K.

Brown F, Griffiths E, Horaud F, Petricciani JC (eds): Safety of Biological Products Prepared from Mammalian Cell Culture.
Dev Biol Stand. Basel, Karger, 1998, vol 93, pp 73-78

Transmission of Spongiform Encephalopathy Through Biological Products

P. Brown

Laboratory of Central Nervous System Studies, National Institute of Neurological Disorders and Stroke, National Institutes of Health, Bethesda, MD, USA

Key word: Transmissible spongiform encephalopathy, transmissible cerebral amyloidosis, kuru, Creutzfeldt-Jakob disease, prion disease.

Abstract: The transmissible spongiform encephalopathies (TSE) are rare but fatal neurological diseases that exist in sporadic, familial, and environmentally acquired forms. Environmentally acquired disease has until now been mostly iatrogenic in origin, recognized as responsible for nearly 200 deaths during the past 20 years. The two most important causes have been contamination of cadaver-derived human growth hormone and dura mater grafts, but a few instances related to contaminated neurosurgical instruments and corneal grafts have also occurred. Currently, a great deal of concern is being voiced about the potential risk of acquiring disease through blood or blood products, and although there is no supporting epidemiological evidence for such a danger, experiments are underway to define which blood components or plasma derivatives might pose the greatest problem.

Spongiform encephalopathy describes a small group of diseases that affect both humans and animals, of which the best known examples are Creutzfeldt-Jakob disease (CJD) in humans, and scrapie and «mad cow disease» (bovine spongiform encephalopathy, or BSE) in animals. The human diseases are characterized clinically by the principal symptoms of dementia and ataxia, with a fatal outcome in most cases of less than one year, and pathologically by neuronal loss, neuropil vacuolation, gliosis, and amyloid deposits consisting of a proteinase-resistant protein (PrP) encoded by a gene on chromosome 20. Most cases are sporadic, but about 10% of cases are familial, associated with any of more than 20 mutations in the chromosome 20 gene. A comparatively small number of cases are iatrogenic, resulting from medical therapeutic misadventures.

Contaminated biological products account for the great majority of such misadventures; indeed, if cooked bodily parts from corpses can be categorized as biologicals, the transmission of kuru through the practice of ritual cannibalism among the Fore population of the Papua New Guinea Highlands may be considered to be the first recognized instance of biological transmission of a human spongiform encephalopathy. A zoonotic alternative seems unlikely, because when the disease first appeared in the early 20th century, sheep and cows did not exist in the region, and pigs and chickens kept by the people are not naturally susceptible to infection.

The most plausible explanation for the spread of kuru, therefore, is that by a twist of fate, a Fore individual dying of CJD, which we now know occurs in a random fashion all over the world at an annual incidence of about one case per million people, was cannibalized, and then propagated the disease throughout the population in successive cannibalistic ceremonies. By mid-century, kuru had achieved epidemic proportions, decimating the most severely affected villages, but following the cessation of cannibalism around 1960 it has now virtually disappeared.

Kuru was shown to be experimentally transmissible to non-human primates in 1966 [1], and within a few years of this discovery came the first report of human-to-human transmission of CJD. In 1974 and again in 1977, isolated instances of iatrogenic CJD were reported; the first in a patient who had received a corneal graft [2], and the second in two patients who had been subjected to stereotactic electroencephalography [3]. Retrospective studies then uncovered a few earlier cases having probably resulted from contaminated neurosurgical instruments. However, iatrogenic disease did not become headline news until 1985, when young hypopituitary patients treated with cadaveric-derived native human growth hormone began dying of CJD [4].

The current tally of such cases stands at 103, mostly in patients residing in the United States, England, and France, although cases have also occurred in Scandinavia, Holland, Brazil, and New Zealand (in the latter two countries from hormone prepared in the United States). The risk of developing CJD is markedly different in the three countries, and is inversely related to the average incubation period elapsing between treatment and the onset of CJD (Table 1). This suggests that the level of contamination differed from country to country, being least in the United States, and greatest in France (the higher the level of contamination, the greater the risk of disease, and the shorter the incubation period). One case that occurred in the United States deserves special mention because of an incubation period of at least 35 years, which instructs us to be cautious in counseling patients about enough time having elapsed since the end of therapy for them to be considered as being out of danger.

Within two years of our recognition of hGH-related CJD in 1985, the first case of iatrogenic disease from contaminated dura mater grafts was identified [5]. The number of such patients has since increased to 80, almost all of whom received

Table 1: Comparative risks and mean incubation periods of Creutzfeldt-Jakob disease (CJD) following treatment with cadaveric human growth hormone grouped according to country in which hormone was prepared.

Country	Number of patients	Treated population	Risk of CJD	Mean incubation period (years)
United States[1]	22	8500	0.3%	18 (± 6)
United Kingdom	26	1750	1.5%	14 (± 3)
France	53	1700	3.1%	8 (± 3)

1 Includes one Brazilian patient and five New Zealand patients who received hormone prepared in the United States.

grafts collected and processed by a single company during the period 1981 through 1985. Until this time, the company had used irradiation to sterilize the dural patches, which was not adequate to inactivate the agent of CJD, but since then the company has added an exposure to 1N NaOH, which is a far more effective decontaminating procedure.

Both the incubation period and clinical presentation of CJD appear to depend upon the route of infection in different types of iatrogenic disease (Table 2). With direct intracerebral or ocular nerve infection (neurosurgical instruments or corneal grafts), incubation times are measured in months (one corneal graft case excepted), and the onset of illness is indistinguishable from that of sporadic cases of CJD. With cerebral surface infection (dura mater grafts) incubation times vary widely from months to years, and the onset of illness may occasionally resemble sporadic disease, but is in most cases predominantly cerebellar. After peripheral routes of infection (pituitary hormone therapy), incubation times are sometimes prolonged for decades, and the onset of illness is uniformly cerebellar [6].

The question of genetic susceptibility to exogenous infection has now been investigated in a sufficiently large number of cases for us to be confident of the results. As with sporadic CJD, iatrogenic disease is overwhelmingly associated with homozygosity at polymorphic codon 129 of the PrP gene [6]. The pathogenetic basis for this effect remains obscure, but has been proposed to involve a process in which similarly configured protein molecules associate to form dimers, polymers and chain-polymer amyloid fibrils.

All known cases of iatrogenic CJD have resulted from exposure to infectious brain, pituitary, or ocular tissue, almost certainly due to the high levels of infectious agent in the central nervous system (and organs to which it is directly connected). However, from tissue distribution studies in both humans and experimental animals, it has been well established that the infectious agent is widespread in the body, albeit in much lower concentration than in the brain, and with an irregular and unpredictable occurrence; thus, there is reason to suppose that at some future time we will learn of a case for which a peripheral tissue is implicated as the source of contamina-

Table 2: Iatrogenic cases of Creutzfeldt-Jakob disease grouped according to the mode of infection (route of entry of the infectious agent into the brain).

Mode of infection	Number of patients	Agent entry into brain	Mean incubation period (range)	Clinical presentation
Corneal transplant	3	Optic nerve	16, 18, 320 mos	Dem/Cereb
Stereotactic EEG	2	Intra-cerebral	18 mos (16, 20)	Dem/Cereb
Neurosurgery	4	Intra-cerebral	20 mos (15-28)	Vis/Dem/Cereb[1]
Dura mater graft	80	Cerebral surface	5.5 yrs (1.5-12)	Cereb (Vis/Dem)
Gonadotrophin	4	Hematogenous	13 yrs (12-16)	Cerebellar
Growth hormone	103	Hematogenous	12 yrs (5-30)	Cerebellar

[1] Vis, visual; dem, dementia; cereb, cerebellar.

tion. We are already aware of a few cases of CJD in patients who two to five years earlier had a pericaridal homograft for tympanic membrane closure, or bone, kidney, or liver transplants; however, in none of these instances was the donor identified as having had CJD, so their iatrogenic origin must still be considered conjectural.

Other biological products are also candidates for evaluation as potential sources of iatrogenic infection. CNS gangliosides of either human or bovine origin have an obvious potential to be contaminated. Other products that derive from animal species not known to be susceptible to spongiform diseases, or even from non-animal sources, may nevertheless be exposed to culture media and protein broths containing small amounts of tissue extracts from a variety of animals, and stabilized with human or bovine albumin in the course of processing protocols. Each such product must be evaluated on an individual basis, and although an informed judgment can usually be made about the potential risk inherent in its components, no amount of reasoning can take the place of «validation» experiments to determine whether in fact infectious starting material retains any infectivity in the final product.

Such studies have been performed only for human growth hormone, bovine gangliosides, and dura mater. Growth hormone remains biologically active after exposure to 6 M urea and ultrafiltration steps that effectively decontaminate the purified hormone [7]. Gangliosides withstand exposure to NaOH and steam autoclaving that totally inactivate the infectivity present in the starting brain material [8]. Dura mater from scrapie-infected hamsters loses virtually all infectivity after treatment with 1N NaOH and retains its morphological integrity [9]. Similar validation studies will be required to determine if the harsh physico-chemical treatments and ultrafiltration steps that are required to disinfect the agent of CJD are compatible with the therapeutic activity of a given biological product.

A final word is required concerning the risk of disease transmission from blood or blood products to which an individual in the pre-clinical or clinical phase of CJD has contributed. Epidemiological studies have not uncovered any instances in which an individual contracted CJD from the blood of any other individual, and have eliminated blood as a statistically important cause of CJD. However, laboratory studies of experimentally infected animals clearly indicate that blood (and particularly buffy coat) may contain low levels of infectivity [10]. Although attempts to isolate infectivity from the blood of humans with CJD are less clear-cut, it is probable that human blood may also be infectious (Table 3) [11-15].

We have recently extended these observations in experiments measuring infectivity in blood components and Cohn plasma fractions from a deliberately «spiked» blood specimen (scrapie-infected hamster brain cells added to human blood), and from the blood of mice experimentally infected with CJD. Infectivity in the «spiked» blood was not limited to the buffy coat, but was also present in red cell and plasma components, and (in decreasing amounts) in all Cohn fractions. Similarly, low levels of endogenous infectivity were present in TSE-infected mouse buffy coat, plasma, and the first two Cohn fractions [16]. We have initiated further experiments in TSE-infected mice to determine the distribution of blood infectivity during the pre-clinical phase of disease, and whether the infectivity is capable of transmitting disease when the blood specimens are assayed intravenously rather than intracerebrally. Parallel experiments are being carried out in primates inoculated with human CJD blood cell and plasma specimens.

In summary, the practice of transferring tissues, tissue extracts, or bodily fluids from one individual to another is never without risk, but the minimal likelihood of contamination by the causative agent of CJD is usually far outweighed by the therapeutic benefit of the product. We must nevertheless be ever vigilant to the ways by which such products can be made even safer, either by donor screening or decontamination procedures, and to be alert to the use of therapeutically acceptable synthetic alternatives.

Table 3: Summary of attempted transmission of human spongiform encephalopathy from blood or blood components of infected patients.

Diagnosis[1]	Inoculated material	Assay animal	Route of inoculation[2]	Pos./total donors	Reference
Spor. CJD	Buffy coat	Guinea pig and hamster	ic	2/2	11
Spor. CJD	Blood	Mouse	ic	1/3	12
Spor. CJD	Plasma	Mouse	ic	1/1	13
Spor. CJD	Blood (units)	Chimpanzee	iv	0/3	14
Spor. CJD	Blood	Spider monkey	ic, iv, ip	0/1	14
Spor. CJD	Blood	Squirrel monkey	ic, ip, im	0/1	14
Spor. CJD	Buffy coat	Squirrel monkey	ic, iv	0/4	14
Spor. CJD	Blood	Guinea pig	ic, ip	0/1	14
Kuru	Serum	Mouse	ic	0/3	14
Iatro. CJD	Blood	Hamster	ic	1/1	15

1 Spor., sporadic; Iatro., iatrogenic.

2 ic, intracerebral; iv, intravenous; ip, intraperitoneal; im, intramuscular.

References

1 Gajdusek DC, Gibbs CJ Jr, Alpers M: Experimental transmission of a kuru-like syndrome in chimpanzees. Nature 1966;209:794-796.

2 Duffy P, Wolf J, Collins G, DeVoe A, Streeten B, Cowen D: Possible person-to-person transmission of Creutzfeldt-Jakob disease. New Engl J Med 1974; 290:692.

3 Bernoulli C, Siegfried J, Baumgartner G, Regli F, Rabinowics T, Gajdusek DC, Gibbs CJ Jr: Danger of accidental person-to-person transmission of Creutzfeldt-Jakob disease by surgery. Lancet 1977;i:478-479.

4 Brown P, Gajdusek DC, Gibbs CJ Jr, Asher DM: Potential epidemic of Creutzfeldt-Jakob disease from human growth hormone therapy. N Engl J Med 1985;313:728-731.

5 Thadani V, Penar PL, Partington J, Kalb R, Jansson R, Schonberger LB, et al: Creutzfeldt-Jakob disease probably acquired from a cadaveric dura mater graft. J Neurosurg 1988;69:766-769.

6 Brown P: Environmental causes of human spongiform encephalopathy, in Baker HF, Ridley RM (eds): Methods in Molecular Medicine: Prion Diseases. Totowa, New Jersey, Humana Press, 1996, pp 139-154.

7 Pocchiari M, Peano S, Conz A, Eshkol A, Maillard F, Brown P, Gibbs C J Jr, Geng Xi Y, Tenham-Fisher E, Macchi G: Combination ultrafiltration and 6 M urea treatment of human growth hormone effectively minimizes risk from potential Creutzfeldt-Jakob disease virus contamination. Horm Res 1991;35:161-166.

8 Di Martino A, Safar J, Ceroni M, Gibbs CJ Jr: Purification of non-infectious ganglioside preparations from scrapie-infected brain tissue. Arch Virol 1992;124:111-121.

9 Diringer H, Braig HR: Infectivity of unconventional viruses in dura mater. Lancet 1989;i:439-440.

10 Brown P: Can Creutzfeldt-Jakob disease be transmitted by transfusion? Curr Opin Hematol 1995;2:472-477.

11 Manuelidis EE, Kim JH, Mericangas JR, Manuelidis L: Transmission to animals of Creutzfeldt-Jakob disease from human blood. Lancet 1985;ii,896-897.

12 Tateishi J: Transmission of Creutzfeldt-Jakob disease from human blood and urine into mice. Lancet 1985;ii:1074.

13 Tamai Y, Kojuma H, Kitajima R, Taguchi F, Ohtani Y, Kawaguchi T, Miura S, Sato M, Ishihara Y: Demonstration of the transmissible agent in tissue from a pregnant woman with Creutzfeldt-Jakob disease. New Engl J Med 1992;327:649.

14 Brown P, Gibbs CJ Jr, Rodgers-Johnson P, Asher DM, Sulima MP, Bacote A, Goldfarb, LG, Gajdusek DC: Human spongiform encephalopathy: the National Institutes of Health series of 300 cases of experimentally transmitted disease. Ann Neurol 1994;35:513-529.

15 Deslys JP, Lasmézas C, Dormont D: Selection of specific strains in iatrogenic Creutzfeldt-Jakob disease. Lancet 1994;343:848-849.

16 Brown P, Rohner RG, Dunstan BC, MacAuley C, Gajdusek DC, Drohan WN: The distribution of infectivity in blood components and plasma derivatives in experimental models of transmissible spongiform encephalopathy. Transfusion, 1998, in press (September issue).

Dr. P. Brown, NIH, Building 36, Room 5B21, Bethesda, MD 20892, USA.

Brown F, Griffiths E, Horaud F, Petricciani JC (eds): Safety of Biological Products Prepared from Mammalian Cell Culture.
Dev Biol Stand. Basel, Karger, 1998, vol 93, pp 79-84

New Variant Creutzfeldt-Jakob Disease

R.G. Will

National Creutzfeldt-Jakob Disease Surveillance Unit, Western General Hospital, Edinburgh, UK

Key word: New variant Creutzfeldt-Jakob disease, bovine spongiform encephalopathy, epidemiology, risk factors.

Abstract: New variant Creutzfeldt-Jakob disease is a novel human spongiform encephalopathy with a consistent clinico-pathological phenotype. Epidemiological evidence indicates that this disease is occurring almost exclusively in the UK, where there has been an epidemic of spongiform encephalopathy in the cattle population. Current evidence strongly supports the hypothesis that there is a causal link between bovine spongiform encephalopathy and new variant Creutzfeldt-Jakob disease.

Introduction

Surveillance of Creutzfeldt-Jakob disease (CJD) was initiated in the UK in May 1990 to identify any change in the characteristics of this condition following the occurrence of bovine spongiform encephalopathy (BSE) in British cattle. In 1996 a novel form of CJD, designated new variant CJD (nvCJD), was identified by the CJD Surveillance Unit and it was proposed that these cases might be causally linked to the occurrence of (BSE). Since the publication of an article describing the cases and the predominantly epidemiological evidence implying a causal link [1], there has been extensive comment and debate on the strength of these data. In this article the overall evidence in relation to the proposed causal link between BSE and nvCJD is reviewed.

Clinico-Pathological Phenotype

The original publication included details of ten cases of an apparently novel form of CJD, designated nvCJD. These cases were first linked because of a remarkably early age of onset (mean 29 years) compared to previous experience of CJD from systematic surveys which consistently indicated that CJD was predominantly a disease of late middle-age. Clinically the cases were also unusual because of a consistent phenotype, including early psychiatric symptoms and a prolonged duration of illness averaging over 12 months in comparison to 4 months in classical cases. Review of case records held at the CJD Surveillance Unit (CJDSU) and of the previous literature indicated that this clinical phenotype was highly unusual in CJD, although rarely there were young patients with classical CJD with similar features. However as a group these cases exhibited a strikingly similar and distinct clinical phenotype consistent with the proposed novelty of this form of CJD.

By August 1997 11 further cases of nvCJD have been identified in the UK and a further case in France [2]; all have a clinical phenotype consistent with the original description. The incidence of CJD in cases aged less than 35 years at death, based on the original 10 cases, was statistically clearly distinct from previous experience. The current situation in the UK is that 15 of the 21 confirmed cases of nvCJD were aged less than 35 years at death and 19 of 21 cases (including one probable case) were aged 35 years or less at disease onset. The duration of illness in the 21 UK cases averages 16 months with a range of 9-38 months. The average age at onset of symptoms in the cases identified in the UK is 27 years with a range of 16-48 years. The year of onset of clinical symptoms and year of death of the cases is listed in Table 1, and indicates no definite change in the rate of occurrence of nvCJD, particularly in relation to disease onsets. The overall evidence is consistent with the continuing occurrence of a novel form of CJD.

The most powerful evidence linking the first 10 cases was the neuropathological phenotype and all subsequent cases have exhibited neuropathological features consistent with the original description. All cases exhibit widespread florid plaque deposition in addition to the spongiform change, astrocytic gliosis and neuronal loss typical of classical CJD. Quantitative image analysis in a sample of cases has confirmed prion protein (PrP) deposition in the brain which exceeds by a factor of two or three the amount of PrP seen in previous cases of classical CJD. There are two possible caveats to this evidence: would the neuropathological features have been identified in the past without modern immunocytochemical techniques and were there cases of CJD with this neuropathological appearance either currently or in the past in countries without BSE? Review of material from past cases of CJD held at the CJDSU had been carried out before announcing of the identification of nvCJD and has continued but no similar past case has been identified. Crucially, extensive review of neuropathological archives in other countries has failed to lead to the identification of even a single similar case. On current evidence the neuropathology of nvCJD is indeed novel, as originally proposed.

Epidemiology

The proposition that BSE and nvCJD were causally linked was based in part on the novelty of nvCJD. However, the identification of a new clinico-pathological form

Table 1: Year of clinical onset and death of nvCJD cases in the UK.

	Onset	Death
1994	7	0
1995	9	3
1996	5	10
1997	0	7 (alive 1)

of CJD does not in itself allow any conclusion regarding aetiology. Variants of CJD have been identified in the past, for example the Heidenhein, Brownell-Oppenheimer and pan-encephalopathic types of CJD. Evidence of a causal link with any putative novel risk factor such as BSE depends on the timing of the occurrence of any related human disease, the geographical co-localisation of such cases with the risk factor and the identification of a common exposure in affected patients.

BSE was identified in 1996 through the UK veterinary surveillance system, although in retrospect there were almost certainly cases in 1985. With an incubation period averaging 5-6 years, exposure of the UK cattle population to infection is likely to have started early in the 1980s [3], presumptively through animal feed in the form of meat and bone meal contaminated initially with sheep scrapie and later with BSE itself. The feeding of ruminant protein to ruminants was banned in July 1988 and the BSE epidemic peaked in 1992/3 and has subsequently significantly declined, consistent with the proposed mechanism of infection. Exposure of the human population to the BSE agent may have started in the early 1980s and may have continued to increase until affected animals were excluded from the human food chain in 1988 and a ban on the use of a range of bovine tissues, including brain and spinal cord from all cattle, in human (and animal) food in 1989. If BSE were a human pathogen, an important question is when the first cases of an associated human disease might develop in relation to the BSE epidemic.

Speculation on the likely timing of the occurrence of a BSE-related human disease is based on inference from data on the incubation period in human prion diseases, including kuru and CJD. In kuru the incubation period ranges from 4.5 to over 30 years, but there are no hard data on the mean incubation period. Evidence on incubation periods in CJD comes from iatrogenic cases [4]. With central inoculation through neurosurgical procedures or dura mater grafts, the incubation periods are relatively short, ranging from 18 months to a few years, although there are occasional cases of dura mater-related CJD with incubation periods exceeding 10 years. Transmission of BSE to humans would not involve direct implantation of infectivity in the CNS but oral exposure through dietary contamination or perhaps some other peripheral route. The closest analogy in iatrogenic CJD is the occurrence of cases following human pituitary hormone therapy administered by peripheral injection. In these cases there is a wide range of incubation periods from about 5 years to over 25 years, with a mean of 13 years. In prion diseases the incubation period is shortest with intra-species transmission and is usually extended when crossing from one species to another. The incubation period of BSE in humans is therefore likely to be longer than in pituitary hormone-related CJD; with human exposure to BSE potentially starting in the early 1980s it is reasonable to propose that a BSE-related human disease might first occur in the mid 1990s, with a putative minimum incubation period ranging from approximately 6-12 years (Fig. 1). The disease onsets of the first cases of nvCJD were in 1994.

If exposure to the agent of BSE has resulted in a human disease, this disease should occur predominantly in the UK as BSE has largely been a disease of British cattle. By October 1996 there had been over 165,000 cases of BSE in the UK, compared to a small number of cases in other countries: 129 in the Republic of Ireland, 224 in Switzerland, 47 in Portugal, 23 in France and cases in single figures in Denmark, Germany, Italy and the Netherlands. Methodologies for CJD surveillance in France, Germany, Italy, The Netherlands, Slovakia and the UK were harmonised in 1993 in order to obtain comparable data and to identify any change in the characteristics of CJD and in particular any change that might be linked to BSE. Detailed

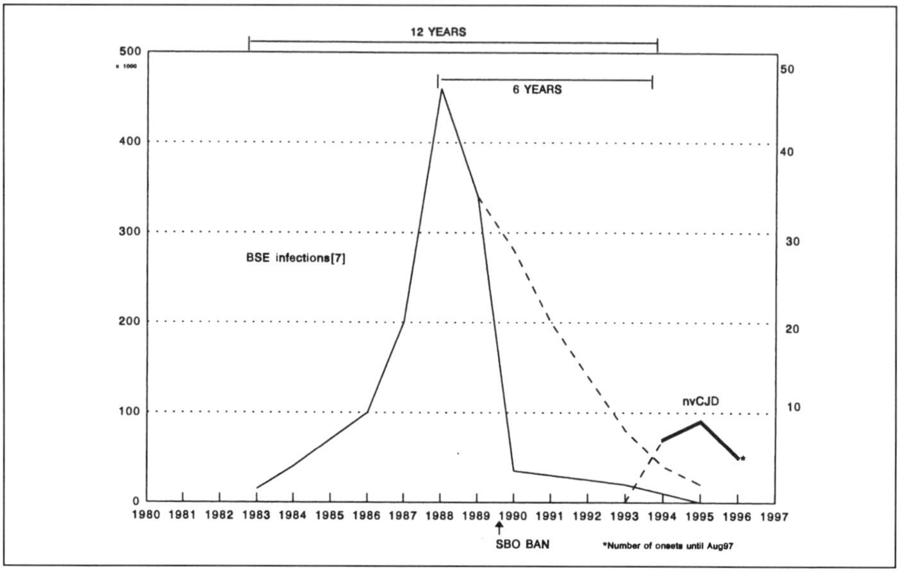

Fig. 1: BSE, «exposure» and incidence of nvCJD (by year of onset).

review of clinical and pathological information in March 1996 indicated that no case of nvCJD had been identified in any collaborating country other than the UK. Since then 11 further cases of nvCJD have been identified in the UK and only one case in continental Europe. This single French case does not refute the hypothetical causal link with BSE as significant quantities of meat products were exported from the UK to France in the 1980s as well as cattle and cattle feed. With effective surveillance of CJD in a large proportion of the continental European population and currently no other cases of nvCJD, it is apparent that nvCJD is occurring almost exclusively in the UK, the country with by far the highest incidence of BSE. This geographical co-localisation of BSE and nvCJD is only circumstantial evidence that the two conditions are linked, but evolving epidemiological evidence reinforces this hypothesis. It would be impossible to interpret the occurrence of nvCJD in the UK without comparative data from other countries.

In the study of CJD in the UK each suspect case referred in life is examined at the referral centre and a standardised questionnaire on potential risk factors is completed by interviewing relatives of the case. Information on past medical history, occupation, and dietary history has been obtained in all cases of nvCJD in order to identify any common linking factor and in particular any common exposure to BSE. There is no past medical history of potential iatrogenic exposure, such as human pituitary hormone therapy, in any case. Some cases have undergone surgical procedures, often minor, but with a similar frequency to age-matched control subjects and there is no obvious link in relation to occupational exposures. After the identification of the first cases of nvCJD, one hypothesis was that these cases might represent genetic forms of CJD associated with mutation of the prion protein gene (PRNP). However, full nucleic acid sequencing of the open reading frame of PRNP has excluded any of the mutations known to be associated with hereditary forms of CJD. All cases tested so far are homozygous for methionine at codon 129 of PRNP, which compares with a similar frequency of this genotype in sporadic CJD [5]. Information from the French cohort of human growth hormone related CJD cases suggests that codon 129 genotype may influence the incubation period with early disease in methionine homozygotes, followed by valine homozygotes and finally heterozygotes [6]. In nvCJD it is possible that cases with genotypes other than methionine/methionine at codon 129 may occur in the future with more prolonged incubation periods.

A case-control study of past dietary history has not demonstrated any dietary factor consumed more frequently by nvCJD cases in comparison to age-matched controls. The failure, so far, to identify a common and distinct dietary factor in nvCJD cases should not be interpreted as refuting a causal link between BSE and nvCJD. There are a number of methodological problems with this type of dietary case-control study. Information on dietary history is necessarily obtained from a surrogate witness, details of past dietary exposures (particularly years in the past) are known to be unreliable, and there is evidence of respondent bias in previous dietary studies in CJD. Exposure to bovine brain and/or spinal cord in human food may have been intermittent, a range of food products may have contained infectivity, and there may have been variation with time in the level of contamination and the products contaminated. It is possible that single meat products only rarely contained sufficient infectivity to initiate agent replication and subsequent disease. The identification of specific causal exposures may be impossible through the retrospective study of dietary histories years later.

In the absence of evidence implicating any specific dietary exposure as the cause of nvCJD, it is important to consider whether there was a population exposure to the BSE agent. A range of tissues from BSE affected animals have been bioassayed in mice to determine the tissue distribution of infectivity and only brain, spinal cord and retina have been found to contain infectivity in natural BSE. Although these experiments cannot exclude low levels of infectivity in apparently negative tissues, bovine CNS contains high titres of infectivity that are probably orders of magnitude higher than any other tissue. If BSE has caused nvCJD the likeliest mechanism is through past dietary exposure to high titres of infectivity through consumption of bovine brain and spinal cord. Currently there is no direct information on past inclusion rates of these tissues in human food products in the UK. However an investigation on the contemporary use of these products has been carried out in Australia and New Zealand, and this may provide some indication of past practices in the UK as dietary habits in these countries are similar to those in the UK and no legislative action has been necessary in these countries which are free of all animal spongiform encephalopathies. With rare exceptions bovine brain does not enter the human food chain in either Australia or New Zealand. A detailed investigation in New Zealand has however indicated that cattle remains, including vertebral column, are used in the production of mechanically recovered meat (MRM). Spinal cord is removed from the majority of carcasses, but spinal cord remnants, including occasional whole cords, are present in a significant proportion of vertebral columns used in the production of MRM. In the UK before 1988, spinal cord from affected cattle may have been included in MRM and, until the introduction of the specified bovine offals ban in November 1989, spinal cord from animals incubating the disease may also have been included. Mathematical modelling has indicated that large numbers of cattle incubating BSE, but clinically unaffected, may have been slaughtered in the late 1980s [7]. It is likely that there was exposure, and potentially extensive exposure, of the population in the UK to CNS titres of BSE infectivity through the use of MRM in the 1980s. This product is said to have been used in a variety of meat products, including burgers, pies, sausages and pâte, but there are no hard data on which products contained MRM, nor on inclusion rates. This type of information would be of particular use in targetting questions on dietary history in nvCJD. However, current evidence indicates that there is likely to have been significant dietary exposure to high titre BSE infectivity in the UK in the 1980s.

Conclusion

Analysis of the clinical and pathological features of nvCJD originally suggested that this was a novel disease. The continuing occurrence of cases with a consistent clinico-pathological phenotype and the failure to identify any similar historical case in the UK or elsewhere provides strong support of the hypothesis that this is a new form of human spongiform encephalopathy. An exposure to the BSE agent in the 1980s and the appearance of a related human disease in the mid 1990s is consistent with the limited available information on incubation periods in human prion diseases. Cases of nvCJD are occurring almost exclusively in the UK, co-localising with the epidemic of BSE. Analysis of risk factors has provided no direct evidence of common specific exposures to BSE in nvCJD, but these may never be identified because of the limitations of retrospective investigation. This evidence is not conclusive, but in the absence of any other explanation, it is probable that BSE and nvCJD are causally linked.

References

1 Will RG, Ironside RG, Zeidler M, Cousens SN, Estibeiro K, Alperovitch A, Poser S, Pocchiari M, Hofman A, Smith PG: A new variant of Creutzfeldt-Jakob disease in the UK. Lancet 1996;347:921-925.

2 Chazot G, Broussole E, Lapras CI, Blattler T, Aguzzi A, Kopp N: New variant of Creutzfeldt-Jakob disease in a 26-year old French man. Lancet 1996;347:1181.

3 Wilesmith JW, Wells GAH, Cranwell MP, Ryan JB: Bovine spongiform encephalopathy: epidemiological studies. Vet Rec 1988;123:638-644.

4 Brown P, Preece MA, Will RG: 'Friendly fire' in medicine: hormones, homografts and Creutzfeldt-Jakob disease. Lancet. 1992;340:24-27.

5 Windl O, Dempster M, Estibeiro JP, Lathe R, de Silva R, Esmonde E, Will R, Sprigbett A, Campbell TA, Sidle KCL, Palmer MS Collinge J: Genetic basis of Creutzfeldt-Jakob disease in the United Kingdom: a systematic analysis of predisposing mutations and allelic variation in the PRNP gene. Hum Gen 1996;98:259-264.

6 Billette de Villemeur T, Deslys J-P, Pradel A, Soubrie C, Alperovitch A, Tardieu M, Chaussain J-L, Hauw J-J, Dormont D, Ruberg M, Agid Y: Creutzfeldt-Jakob disease from contaminated growth hormone extracts in France. Neurology 1996;47:690-695.

7 Anderson RM, Donnelly CA, Ferguson NM, Woolhouse MEJ, Watt CJ, Udy HJ, MaWhinney S, Dunstan SP, Southwood TRE, Wilesmith JW, Ryan JBM, Hoinville LJ, Hillerton JE, Austin AR, Wells GAH: Transmission dynamics and epidemiology of BSE in British cattle. Nature 1996;382:779-788.

Dr. R.G. Will, CJD Surveillance Unit, Western General Hospital, Edinburgh EH4 2XU, UK.

Brown F, Griffiths E, Horaud F, Petricciani JC (eds): Safety of Biological Products Prepared from Mammalian Cell Culture.
Dev Biol Stand. Basel, Karger, 1998, vol 93, pp 85-88

Problems with BHK 21 Cells

F. Brown

Plum Island Animal Disease Center, Greenport N.Y., U.S.A.

Key word: Foot-and-mouth disease virus; BHK cells.

Abstract: The properties of baby hamster kidney (BHK 21) cells are modified by passage in suspension culture. The suspended cells differ from monolayer cells in the surface expression of some integrin chains involved in attachment of foot-and-mouth disease virus (FMDV), in particular the progressive down-regulation of both $\alpha5$ and αV integrin chains. This down-regulation is correlated with the loss of actin stress fibres. FMDV particles from these cells are unstable towards the aziridine used in inactivating the virus for vaccine production. Moreover, growth of virus in suspended cells can lead to the selection of antigenic variants which differ from those produced in monolayer cells.

Introduction

Baby hamster kidney (BHK C 13) cells are a continuous line of hamster kidney cells derived by MacPherson and Stoker in 1962 [1]. The original isolates were pseudodiploid and anchorage-dependent, and possessed the property of an indefinite life in culture by serial sub-cultivation. Mowat and Chapman [2] showed that the cells produced high levels of FMDV and for this reason were studied intensively because of their potential for the large-scale production of virus required for vaccine production. On the basis of many years experience it is safe to conclude that most FMDV isolates can be adapted to levels of growth in BHK monolayers which are sufficient for vaccine production and several commercial enterprises have used this method for vaccine production.

Nevertheless, the anchorage-dependent characteristics place a considerable limitation on the potential for scale-up. This persuaded Capstick et al [3] to investigate whether the cell could be adapted to growth in suspension culture. Aneuploid sublines were derived from the parent anchorage-dependent cells and these are used for the production of a large proportion of the FMD vaccines which are manufactured today. Despite the early concern that these suspension cells may be tumorigenic, extensive experiments have failed to reveal any such problems. This issue was reviewed by Regan and Petricciani at a meeting on the same topic, held in November 1986 [4]. They concluded that, although the intact cells are tumorigenic in hamsters, the species of origin of the cell line, no such activity has been demonstrated in a variety of other species. Even in hamsters, the cells must be intact to cause tumour growth. This situation can be avoided readily in the production of biologicals.

Current problems

The problems that have arisen in the use of BHK cells are connected with the differing characteristics that arise in culture. It has not been established whether these have arisen by selection from a heterogeneous parent population or by mutation. The chief concern in the use of the cells for the production of FMDV has been the lack of uniformity regarding susceptibility. No single cell sub-line appears to be universally more susceptible or productive than another. These problems have been studied recently by Amadori and his colleagues [5]. Their studies showed striking genetic and phenotypic changes in the cells in suspension culture. Alteration from diploid to aneuploid during adaptation had been reported in 1966 by Capstick et al [6]. A further alteration towards a high prevalence of polyploid cells can also take place after repeated passages. It seems likely that these have been selected from an original heterogeneous population which contains distinct sub-populations [7, 8].

After several passages in suspension culture, the cells had a much reduced plating efficiency and a low surface expression of the $\alpha5\ \beta1$ integrin, despite its high level of synthesis. This has also been observed in recent work on transformed rodent fibroblasts [9]. These changes have important implications for the growth of FMDV. The first of these is the much reduced yields of virus which are found with some isolates. Amadori and his colleagues have identified several reasons for this observation [5].

Attachment of virus to cells

Attachment of FMDV to susceptible cells is principally. but not entirely, via the arginine, glycine, aspartic acid (RGD) triplet situated on the prominent GH loop of VP1, one of the capsid proteins [10, 11, 12]. RGD-specific integrins are present on BHK cells; in particular $\alpha V\beta3$ and $\alpha5\beta1$ have been implicated in attachment [5, 12, 13]. Reduction in surface expression of the $\alpha5\beta1$ integrin would almost certainly result in reduced yields of virus. Moreover, because more than one integrin is involved in cell attachment, this reduction in $\alpha5\beta1$ integrin could lead to selection within the virus population (see below).

Morphogenesis of virus particles

Reduced yields of virus could also be related to impaired morphogenesis of virus particles. Internal membranes play a crucial role in the replication of picornavirus RNA and capsid assembly [5]. Amadori and his colleagues have speculated that FMDV needs a non-membranous sub-stratrum such as microfilaments to transfer the genome-linked protein VPg to the RNA. This function could be accomplished by actin, which is also present in virus particles [14]. In the absence or defective organisation of microfilaments, the balance between RNA production and transfer of VPg would be impaired and would lead to an abnormal accumulation of empty capsids. This latter event has been observed.

Interestingly, the virus particles produced in suspended cells are more labile towards binary ethyleneimine (the inactivant used in FMD vaccine production) than those produced in monolayer cells. Integrity of the virus during BEI treatment was correlated with the distribution of organised actin in the cells. Poor integrity was found with virus grown in cells with intra-cytoplasmic stacks of organised actin but no stress fibres. This resulted in vaccines which failed vaccine potency tests.

Antigenic variability arising from passage in suspended cells

More than 20 years ago, Cowan et al [15] and Meloen [16] showed that the antigenic character of FMDV was changed by passage in different strains of BHK cells. These observations were extended by Bolwell et al [17] who found that vaccine prepared from a virus of serotype A (A22, Iraq 24/64) adapted to grow in BHK monolayer cells induced antibodies which neutralized many isolates belonging to the A serotype. However, vaccine prepared from a virus obtained by passage in BHK-suspended cells of the monolayer cell-adapted virus induced antibodies with highly specific neutralizing activity. In addition to their differing antigenic properties, the two viruses could be distinguished by plaque morphology, their differing ability to attach to monolayer cells, and a tendency for the suspension cell-adapted virus to aggregate. Sequencing studies revealed three changes in the amino acid composition of VP2, one of the capsid proteins. Surprisingly no changes were found in VP1, the protein carrying the major cell attachment site and the immunodominant epitope.

Conclusions

The observations described in this short review make it clear that careful control of the master BHK cell stock is essential for the large-scale production of FMDV in suspension culture. Not only are the surface characteristics of the cells altered by passage in suspension culture but the yield and quality of the virus can also be changed. Moreover, the change in the antigenic properties of the virus which may occur require that the product be carefully monitored so that the product matches the field strain.

References

1 MacPherson I, Stoker M: Polyoma transformation of hamster cell clones - an investigation of genetic factors affecting cell competence. Virology 1962;16:147-151.

2 Mowat GN, Chapman WG: Growth of foot-and-mouth disease virus in a fibroblastic cell line derived from hamster kidneys. Nature 1962;194:253-255.

3 Capstick PB, Telling RC, Chapman WG, Stewart DL: Growth of a cloned strain of hamster kidney cells in suspended cultures and their susceptbility to the virus of foot-and-mouth disease. Nature 1962;195:1163-1164.

4 Regan PJ, Petricciani JC: The approach used to establish the safety of veterinary vaccines produced in the BHK 21 cell line, in Petricciani JC, Hennessen W (eds): Cells, Products, Safety. Dev Biol Stand. Basel, Karger, 1987, vol 68, pp 19-25.

5 Amadori M, Berneri C, Archetti IL: Immunogenicity of foot-and-mouth disease virus grown in BHK-21 suspension cells. Correlation with cell ploidy alterations and abnormal expression of the α5β1 integrin. Vaccine 1994;12:159-166.

6 Capstick PB, Garland AJ, Masters RC, Chapman WG: Some functional and morphological alterations occurring during and after the adaptation of BHK 21 clone 13 cells to suspension culture. Exp Cell Res 1966;44:119-128.

7 Montagnier L, Lage-Davila A: Biological and biochemical criteria for the characterization of the transformed stage of BHK-21 cells, in Regamey RH, Spier R, Horaud F (eds): Second General Meeting of European Society of Animal Cell Technology. Dev Biol Stand. Basel, Karger, 1979, vol 42, pp 27-31.

8 Clarke JB, Spier RE: Variation in the susceptibility of BHK populations and cloned cell lines to three strains of foot-and-mouth disease virus. Arch Virol 1980;63:1-9.

9 Dalton SL, Marcantonio EE, Assoian RD: Cell attachment controls fibronectin and α5β1 integrin levels in fibroblasts. Implications for anchorage-dependent and -independent growth. J Biol Chem 1992;267:8186-8191.

10 Fox G, Parry NR, Barnett PV, McGinn B, Rowlands DJ, Brown F: The cell attachment site on foot-and-mouth disease virus includes the amino acid sequence RGD (arginine-glycine-aspartic acid). J Gen Virol 1989;70:625-637.

11 Baxt B, Becker Y: The effect of peptides containing the arginine-glycine-aspartic acid sequence on the adsorption of foot-and-mouth disease virus to tissue culture cells. Virus Genes 1990;4:73-83.

12 Berinstein A, Roivainen M, Hovi T, Mason PW, Baxt B: Antibodies to the vitronectin receptor (integrin αVβ3) inhibit binding and infection of foot-and-mouth disease virus to cultured cells. J Virol 1995;69:2664-2666.

13 Chambers M, Dougan G, Newman J, Brown F, Crowther J, Mould AP, Humphries MJ, Francis MJ, Clarke B, Brown AL, Rowlands D: Chimeric hepatitis B virus core particles as probes for studying peptide-integrin interactions. J Virol 1996;70:4045-4052.

14 Grigera PR, Tisminetzky SG, Lebendiker MB, Periolo OH, La Torre JL: Presence of a 43-kDa host-cell polypeptide in purified aphthovirions. Virology 1988;165:584-588.

15 Cowan KM, Erol N, Whiteland AP: Heterogeneity of type Asia-1 foot-and-mouth disease virus and BHK-21 cells and the relationship to vaccine preparation. Bull Off Int Epiz 1974;81:1271-1298.

16 Meloen RH: Localisation on foot-and-mouth disease virus (FMDV) of an antigenic deficiency induced by passage in BHK cells. Arch Virol 1976;51:299-306.

17 Bolwell C, Brown AL, Barnett PV, Campbell RO, Clarke BE, Parry NR, Ouldridge EJ, Brown F, Rowlands DJ: Host-cell selection of antigenic variants of foot-and-mouth disease virus. J Gen Virol 1989;70:45-57.

Prof. F. Brown, Plum Island Animal Disease Center, P.O. Box 848, Greenport, N.Y. 11944, USA.

Brown F, Griffiths E, Horaud F, Petricciani JC (eds): Safety of Biological Products Prepared from Mammalian Cell Culture.
Dev Biol Stand. Basel, Karger, 1998, vol 93, pp 89-96

Experiences of Virus, Retrovirus and Retrovirus-like Particles in Chinese Hamster Ovary (CHO) and Hybridoma Cells Used for Production of Protein Therapeutics

S.R. Adamson

Mammalian and Microbial Cell Sciences and Pilot Operations, Genetics Institute, Andover MA, USA

Key word: Virus, retrovirus, CHO cells, hybridoma cells, product safety.

Introduction

In 1991, W.A. Moore [1] made a presentation at the International Workshop on Current Issues (Bethesda, MD) in which he described the cumulative experience of MA Bioservices (formerly Microbiological Associates) in testing mammalian cell cultures for adventitious virus contaminants over approximately a 10-year period. In tests on 550 cell lines, 58% of which were hybridomas and 42% rodent cell lines, the vast majority of which were CHO cells, less than 1% of the cell lines tested were seen to contain adventitious viruses. The conclusion from this survey was that the incidence of adventitious viruses in cell lines used mostly in the biotechnology industry is very low. Notably, results from lot release or end-of-process testing were excluded from this study. In a subsequent PhRMA sponsored survey presented by A.S. Lubiniecki at the International Scientific Conference on Viral Safety and Evaluation of Viral Clearance from Pharmaceutical Products (Bethesda, MD) [2], 100-110 companies were canvassed worldwide and 26 responses were received. In all 70 cell lines, resulting in 88 Master Cell Banks (MCBs) and 126 working cell banks (WCBs), were tested. All the 197 test materials from these cell banks were shown to be free from adventitious viruses. In an extension of the previous survey [1] 516 test articles from unprocessed bulk harvests or end of process samples were shown to be free from contaminating adventitious viruses. Although the conclusion from this survey was clearly that no incidence of viral contamination was reported, the caveat to the survey was that the response was incomplete. Indeed, at the same meeting additional experience was described by R. Garnick, from Genentech [3], who described two independent events in which minute virus of mice (MVM) contamination of a CHO production cell line was observed. The source of contamination was unresolved; however, non-homogeneous contamination of raw materials by mouse faeces or urine were suspected. In addition D. Burstyn [4] representing Biogen, described contamination of a CHO production cell line by epizootic haemorrhagic disease virus (EHDV). The suspected source of contamination was FBS used in the production process. This observation had been previously documented by Rabenau et al (1993). Finally, B. Hughes [5], with Wellcome Laboratories, reported a parvovirus contamination event in a baby hamster kidney (BHK) production cell line in which the suspected source of contamination was serum used in the production process [6].

The purpose of this paper is to extend the previous observations by describing cumulative experience over the last twelve years at Genetics Institute (GI) and to supplement this information with experience from a number of contract laboratories carrying out testing of cell lines for the biotechnology industry. The paper will focus on the cumulative experience of adventitious viruses, retrovirus and retrovirus-like particles (RVLPs) in CHO and hybridoma cells. The survey is, for the most part, restricted to CHO and hybridoma or plasmacytoma cells, since these are the most widely used cells for the production of protein therapeutics within the biotechnology industry.

Results

GI experience in screening for adventitious viral contaminants

Over the last 12 years GI has made many batches of protein products for clinical and commercial use. The cell culture process we have most often used is a continuous process of finite duration, in which production cells (exclusively CHO) are thawed from a WCB and grown for a pre-defined number of generations from the cell bank. Product is harvested from actively growing cells every two-to-four days during this period [7]. Approximately 17 batches of drug substance are produced from the cells during this period (up to approximately 60 generations from the WCB). A battery of virus detection tests, consistent with current points to consider documentation [8] are carried out on MCB/WCB and on cells or unprocessed bulk from the last batch of a manufacturing inoculum to ensure that all the product made during that inoculum was free from virus contamination. In the last 12-year period we have screened 17 cell banks using in vitro virus detection tests employing relevant indicator cell lines and 16 cell banks by in vivo tests using eggs, mice and guinea pigs without detection of adventitious viruses. In addition we have carried out end-of-process testing on 156 samples by in vitro tests and 62 by in vivo tests without viral detection. The total number of harvests covered by this testing plan is 2637 and the total volume of cell culture medium was approximately 5 million litres, 95% of which were serum-free.

Experience in the biotechnology industry in screening for adventitious viral contaminants

To extend the frame of reference to include hybridoma cell lines as well as incorporating additional experience with CHO cells a survey of a number of cell line testing companies and other biotechnology companies was undertaken. Cumulative experience with murine plasmacytoma and hybridoma cell lines used between 1986 and 1996 at Lonza Biologicals (formerly Celltech Biologicals) and tested at Q-One Biotech have been described [9]. Forty-four cell lines were tested by in vitro tests, using MRC-5, Vero and other cell lines. The test articles from these cell lines consisted of 125 fermenter samples and 74 cell bank samples. No adventitious viruses were detected in any of the test materials. Similarly 16 cell lines consisting of one fermenter sample and 27 cell bank samples were tested in vivo without any detection of adventitious viruses. At the Biological Research Faculty and Facility (BRFF, Rockville, MD) a number of different cell lines were tested between 1993 and 1996, of which 80-85% were hybridoma cell lines (the majority of the other cell lines were CHO cells). In vitro tests were carried out on 35 fermenter samples and 35 cell bank samples without detecting adventitious viruses and 49 tests were carried out on cell bank samples with the same result.

Table 1 describes the tests by MA Bioservices (formerly Microbiological Associates, Rockville, MD) between 1992-1996 (prior experience at MA is covered in Moore, 1992) and Table 2 describes the tests by Quality Biotech (Camden, New Jersey) between 1990 and 1996. The experience of both companies was that the frequency of occurrence of adventitious viruses was very low. Notably, in neither case were viral contaminants seen in CHO cells.

Table 1: Survey of in vitro and in vivo tests from MA bioservices, 1992-1996.

Total # of studies: 2014		
Types of cell lines	**Approximate % of total**	
CHO	45-55	
Hybridoma	35-45	
BHK_{21}	5-10	
Other	1-5	
Stage of cell line	**Approximate % of total**	
Cell banks	40	
End-of-process	60	
Test	**Number of positives (%)**	**Number of studies**
In vitro	6 (0.44)	1,366
In vivo	0	648

Observations:

- No adventitious viruses seen in CHO cells;
- Viruses seen in primary bovine cells, BHK_{21}, mouse/human heterohybridoma – all end-of-process samples;
  Viruses unidentified.

Table 2: Survey of in vitro and in vivo tests from Quality Biotech, 1990-1996.

Types of cell lines	**Approximate % of total**	
In vitro		
CHO	33	
Hybridoma	67	
In vivo		
CHO	50	
Hybridoma	50	
Stage of cell line	**Approximate % of total**	
Cell banks	25	
End-of-process	75	
Test	**% positive**	**Cell line**
In vitro	0	CHO
	4.8	Hybridoma
In vivo	0	CHO
	0	Hybridoma

Observations:
Multiple positive samples were from the same mouse/human heterohybridoma and were suspected human CMV. Remainder were positive on haemadsorption – viruses were unidentified.

Retroviruses and retrovirus-like particles in hybridoma and CHO cell lines

Both the published literature and GI's experience have shown the presence of retrovirus-like particles (RVLPs) in CHO cells without evidence of infectivity [10-13]. More recently, a survey of co-cultivation tests carried out at MA Bioservices between 1992 and 1996 shows the same result. In these experiments CHO cells were co-cultivated for approximately five passages with either human rhabdomyosarcoma (RD), mink lung, MRC-5 or human lung carcinoma (A549) cell lines. At the conclusion of this passage cycle, conditioned medium was evaluated by S+L- and reverse transcriptase assays. From 250 samples analysed no evidence of infectious, replicating retrovirus was seen. By transmission electron microscopy (TEM) A-type particles located within the cytoplasm, often in association with centrioles and C-type RLPs, have been observed. Little is known about the A-type particles. However, their intracellular location and low level of budding suggest they are analogous to intracisternal A-type particles (IAPs) found in other rodent cell lines [14] which are intracellular, lack a typical retroviral envelope and are non-infectious [15]. An explanation for the lack of infectivity of CHO C-type RVLPs comes from recent experiments carried out at Genentech using a recombinant CHO cell line [16]. A partial DNA clone of the CHO C-type RVLP was isolated which spanned the 3' end of the Pol gene covering most of the coding sequence for the endonuclease region. The DNA sequence of this clone contained multiple interruptions of potential coding sequences in all three open reading frames, rendering the retrovirus incapable of encoding an intact endonuclease. Typically, between 10^3 and 10^6 RVLPs/mL are present in conditioned medium from industrial cell culture processes (GI experience; personal communications A.S. Lubiniecki, M. Dinowitz and A. Losikoff) as established by TEM of concentrated, conditioned cell culture medium.

Murine plasmacytoma and hybridoma cells are known to produce IAPs and C-type particles. Unlike CHO C-type RVLPs, Hybridoma C-type particles have the ability to replicate in several different cell lines including a small number of human cell lines [1, 17, 18]. However these observations are not consistent with subsequent studies presented to the Vaccines and Related Biological Products Advisory Committee (August 21st 1990, Bethesda MD) by representatives from Centocor, Cetus, IDEC, Organon Technica, Sandoz and Teijin. In these studies all the mouse/human heterohybridomas described showed C-type particles on TEM and manganese-dependent reverse transcriptase activity. However, although most of the cell lines described showed infective xenotropic retrovirus in S+L- assays, the titres of which could be increased by chemical induction or co-cultivation by mink cells, on cultivation with human or primate cells, no infection was observed; the cell lines used in the co-cultivation studies were MRC-5, Vero, human peripheral blood lymphocytes and T cells. In addition, tests carried out at Q-one Biotech on mouse plasmacytoma derived cell lines used at Lonza Biotech between 1986 and 1996 are shown in Table 2 [9]. Although evidence for the presence of infectious xenotropic retrovirus in hybridoma cells is clear (notably 85% of the cell bank samples tested positive by extended S+L-) none of the samples co-cultivated with a human cell line (MRC-5) showed evidence of replicating retrovirus. Quantification of retrovirus-like particles in hybridoma harvest medium showed between 1 and 87 x 10^6 RVLP/mL, approximately 1 in 10^6 of which were shown to be infective in mink S+L- assays [9].

Table 3: Test for mouse cell lines used between 1986 and 1996 at Celltech/Q-One Biotech.

Test	**Number of cell lines tested**	**Percentage of cells positive**	
		Fermenter samples	**Cell bank samples**
Electron Microscopy	22	100% (n = 44)	100% (n = 45)
Reverse transcriptase	45	87% (n = 52)	67% (n = 169)
XC (direct)	24	5% (n = 132)	6% (n = 28)
XC (extended)	18	0% (n = 25)	6% (n = 38)
S + L (direct)	58	28% (n = 17)	42% (n = 65)
S + L (extended)	21	38% (n = 25)	85% (n = 41)
Co-cultivation with human cell line	18	not tested	0% (n = 25)

From Froud et al [9].

Summary and Conclusions

Garnick and coworkers indicated that they experienced two independent MVM outbreaks in a period where approximately 2000 fermentations were performed, hypothesizing that such events were rare but inevitable consequences of very large scale operations [2, 3]. In GIs experience over the last 12 years we have seen no incidence of MVM (or any other virus) in close to 3000 fermentations, albeit at lower volumes than produced at Genentech; GI has used 250-2500L bioreactors for manufacturing whereas Genentech have reported using 100-10,000L bioreactors [3]. Nonetheless, volumes of complex media in the same range as used at Genentech have been used at GI with no observations of viral contamination events. The reason for this is not clear. However, GI's experience in combination with experience from sub-contract testing agencies who service the majority of the biotechnology industry may call the inevitability of an MVM outbreak into question. It would appear that very few adventitious viral contaminations of cell cultures have occurred in industry in the last decade. Interestingly, the frequency of contamination events appear to be lower in CHO cells than in hybridoma cells. It should be noted, however, that these conclusions are not statistically based and the scope of the above survey was somewhat limited.

RVLPs are present in both CHO and hybridoma cells. The characteristics of both are compared in Table 4. C-type particles from hybridoma cells are more abundant as a rule than those from CHO cells. Although the majority of C-type particles produced by hybridoma cells appear to be non-infective (in S+L- assays), approximately one in a million particles are competent to replicate in S+L- cells [9]. The evidence that C-type particles can replicate in human cells has proved difficult to reproduce consistently. It is likely that replication of xenotropic hybridoma C-type particles in human cells is inefficient or restricted to only a small number of specific cell lines. C-type RVLPs from CHO cells are produced less abundantly than those from hybridoma cells and are not competent to replicate due to a defective endonuclease gene [16]. However, over the last two decades the use of hybridoma cells and products derived from these cells has not provided any evidence of transmission of these viruses to humans; in addition they can be readily removed or inactivated. Thus, neither agent would appear to constitute a significant risk to pharmaceutical products made from their respective host cells. Nonetheless, given the difference in relative safety profiles between RVLPs from CHO and hybridoma cells it is not unreasonable to propose that safety factors (clearance factors in

Table 4: RVLPs in CHO and hybridomas.

	Characteristics	
	CHO	**Hb**
A-type	Intracytoplasmic	Intracisternal
C-type	$<10^3$-10^6/mL	$\sim10^6 - \sim 10^8$/mL
	Non-infectious	Infectious – low levels of plaque/foci forming units/mL in S+L-. Infective in human embryonic lung fibroblasts and RD cells. Non-infective in a number of other human and primate cells.

removal/inactivation studies in excess of the reduction of virus loads to zero) required should be less for a CHO process than for a hybridoma process.

Acknowledgements

The author would like to gratefully acknowledge the significant contributions of Tim Charlebois and Barbara O'Connell from Genetics Institute, Dominique Vacante, Anton Steuer and colleagues from MA Bioservices, Steve Froud and colleagues from Lonza and Q-One Biotech, Joe Hughes from Quality Biotech, Andy Losikoff from BRFF, Tony Lubiniecki from Smith Kline-Beecham, and Rob Garnick and Marshal Dinowitz from Genentech.

References

1 Moore WA: Experience in cell line testing, in Brown F, Esber E, Williams M (eds): Continuous Cell Lines. Dev Biol Stand. Basel, Karger, 1992, vol 76, pp 51-56.

2 Lubiniecki AS: Survey of manufacturers and developers of cell culture-derived products for evidence of adventitious contamination, in Brown F, Lubiniecki AS (eds): Viral Safety and Evaluation of Viral Clearance from Biopharmaceutical Products. Dev Biol Stand. Basel, Karger, 1996, vol 88, pp 195-197.

3 Garnick RL: Experience with viral contamination in cell culture, in Brown F, Lubiniecki AS (eds): Viral Safety and Evaluation of Viral Clearance from Biopharmaceutical Products. Dev Biol Stand. Basel, Karger, 1996, vol 88, pp 49-56.

4 Burstyn DG: Contamination of genetically engineered Chinese hamster ovary cells, in Brown F, Lubiniecki AS (eds): Viral Safety and Evaluation of Viral Clearance from Biopharmaceutical Products. Dev Biol Stand. Basel, Karger, 1996, vol 88, pp 199-203.

5 Hughes B: Key points from Nettleton and Rweyemamu (1980): Association of calf serum with parvovirus infection, in Brown F, Lubiniecki AS (eds): Viral Safety and Evaluation of Viral Clearance from Biopharmaceutical Products. Dev Biol Stand. Basel, Karger, 1996, vol 88, pp 205-206.

6 Nettleton PF, Rweyemamu MM: Arch Virol 1980;64:359-374.

7 Adamson SR: Design and operation of a recombinant mammalian cell manufacturing process for rFVIII. Ann Hematol 1994;68:9-14.

8 Points to Consider in the Characterization of Cell Lines Used to Produce Biologicals, Office of Biologics Research and Review, Center for Drugs and Biologics, FDA (USA), 1993.

9 Froud SJ, Birch J, McLean C, Shepherd A, Smith K: Contaminants found in mouse cell lines used in the production of biological products, in Corrondo and Griffiths (eds): ESACT Proceedings. Dordrecht, Kluwer Academic Press, 1997.

10 Lieber MM, Benveniste RE, Livingstone DM, Todaro GJ: Mammalian cell in culture frequently release type C viruses. Science 1973;182:56-59.

11 Tihon C, Green M: Cyclic AMP-amplified replication of RNA tumor virus-like particles in Chinese hamster ovary cells. Nature New Biol 1973;244:227-231.

12 Heine UI, Kramarsky B, Wendel E, Suskind RG: Enhanced proliferation of endogenous virus in Chinese hamster cells associated with microtubules and the mitotic apparatus of the host cell. J Gen Virol 1979;44:45-55.

13 Lubiniecki AS, May LH: Introduction to regulatory practice in novel biotechnology, in Lubiniecki AS, Vargo (eds): Regulatory Practice for Pharmaceutical Production. Wiley-Liss, NY. 1985, pp 1-12.

14 Dalton AJ, Potter M, Merwin RM: Some ultrastructural characteristics of a series of primary and transplanted plasma-cell tumors of the mouse. J Natl Cancer Inst 1961;26:1221-1267.

15 Markus-Sekura CJ, Klutch M, Lundquist M, Dunlap R: Increased expression of CHO cell endogenous retrovirus-like particles detected by electron microscopy after treatment with viral inducing agents or cytokines. In Vitro Toxicology 1991;4:13-25.

16 Anderson KP, Low ML, Lie YS, Keller G-A, Dinowitz M: Endogenous origin of defective retroviruslike particles from a recombinant Chinese hamster ovary cell line. Virology 1991;181:305-311.

17 Spriggs DR, Krueger RG: Identification of the major structural proteins of two BALB/c myeloma C type viruses. Virology 1979;98:35-44.

18 Weiss RA: Retroviruses produced by hybridomas. N Engl J Med 1982;307:1587.

Dr. R. Adamson, Mammalian and Microbial Cell Sciences and Pilot Operations, Genetics Institute, 1 Burtt Road, Andover, MA 01824, USA.

Brown F, Griffiths E, Horaud F, Petricciani JC (eds): Safety of Biological Products Prepared from Mammalian Cell Cülture.
Dev Biol Stand. Basel, Karger, 1998, vol 93, pp 97-107

Human Diploid Cell Strains (HDCS) Viral Vaccines

M.A. Fletcher[1], *L. Hessel*[2], *S.A. Plotkin*[1]

[1] Pasteur Mérieux Connaught, Marnes La Coquette, France
[2] Pasteur Mérieux Connaught, Lyon, France

Key word: Human diploid cell strain (HDCS), WI-38, MRC-5, history, viral vaccines, adverse events.

Abstract: Since the development in 1961 by Hayflick and Moorhead of a human diploid cell strain (HDCS) culture system for isolating viruses, HDCS-derived human vaccines have been licensed worldwide for polio IPV and OPV (multiple strains), rabies (Pitman-Moore L503 3M strain); rubella (RA27/3 strain); measles (Edmonston-Zagreb strain); varicella-zoster (Oka strain); mumps (Rubini strain) and hepatitis A (HM-175, CR-326-F', and GBM strains). Many of these widely-used vaccines now have at least a 20-year record of safety after extensive, ongoing pharmacovigilance. Serious vaccine-associated events are rare, and appear to reflect the activity of the live viruses replicated in the HDCS or of the inactivated viruses or other proteins added during manufacture. There is no credible association of reactions to the HDCS substrate or a hypothetical contaminant derived from it.

History

Until 1961, virus cultures for the production of licensed vaccines were produced in embryonated eggs or through disaggregation of primary animal tissues, which were used with relatively little passage or subcultivation [1]. This changed when Hayflick and Moorhead published the first description of human diploid cell strains (HDCS), which had finite population doublings and which could be used as a culture system for isolating and growing viruses [2]. The strain that was most characterized, called WI-38, was stocked in low passage for vaccine manufacture. WI-38 was derived from foetal lung, which appears to be the easiest tissue from which to grow HDCS. This human cell culture system for viruses was accepted quickly by medical scientists in many parts of the world, and HDCS vaccines soon went into clinical trials: adenovirus vaccine [3] and oral poliovirus vaccine trials in 1963 [4], measles vaccine clinical trials in 1964 [5], and the first rubella vaccine trials in 1965 [6]. The Cell Culture Committee, Permanent Section for Microbiological Standardization, charged with educating scientists about WI-38, had its first meeting in 1963 (Opatija, former-Yugoslavia), where a system of standardized master cell banks, working cell banks, and documentation on «Minimum Requirements» for HDCS was established [7]. In 1969, an important backup HDCS, MRC-5, was developed by Jacobs, Jones and Baille [8].

Some of the important moments in the history of HDCS are outlined in Table 1. An OPV was the first HDCS-derived vaccine to be licensed (1967, former-Yugoslavia) [1]. Subsequently, OPV grown in HDCS was licensed in the UK (1970) and in the USA (1972) [9]. This was followed, in the USA for example, by other HDCS-derived viral vaccines, such as those against rabies (Pitman-Moorc L503 3M strain), rubella (RA27/3 strain), varicella-zoster (Oka strain), and hepatitis A (HM-175, CR-326-F', and GBM strains). Other HDCS vaccines were licensed, such as a measles (Edmonston-Zagreb strain) in France, and a mumps (Rubini strain) vaccine in Switzerland and Italy.

Table 1: A chronology of the evolution of HDCS vaccines.

Year	Event
1961	WI-38 (Hayflick and Moorhead)
1962	NIH discusses use of cell strains and cell lines for human virus vaccines (USA) WI-38 patent application denied (USA)
1963	Adenovirus vaccine clinical trials (Bierly; USA) OPV vaccine clinical trials (Ikic; former-Yugoslavia) First Cell Culture Committee Conference (Opatija, former-Yugoslavia)
1964	Measles vaccine clinical trials (Andzaparidze; Russia)
1965	Rubella vaccine clinical trials (Plotkin; USA)
1967	Division of Biologic Standards begins developing non-human diploid cell lines (USA) OPV vaccine licensed (former-Yugoslavia)
1969	MRC-5 (Jacobs, Jones, and Baille) Rubella vaccine in HDCS criticised by FDA (USA)
1970	Rubella vaccine (RA27/3) licensed (UK, France) OPV vaccine licensed (UK)
1972	OPV vaccine licensed (USA)
1979	Rubella vaccine (RA27/3) licensed (USA)
1980	Rabies vaccine (Pittman-Moore L503 3M) licensed (USA)
1986	Mumps vaccine (Rubini) developed (Switzerland)
1987	Varicella vaccine (Oka) licensed (Japan)
1992	Hepatitis A vaccine (HM-175) licensed (Belgium)
1995	Varicella vaccine (Oka) licensed (USA)
1996	Hepatitis A vaccine (CR-326 F') licensed (USA)

Advantages and Disadvantages of HDCS

For vaccine production, HDCS cells possess a number of favourable qualities. They are susceptible to many human viruses; they can be well characterized; the early population doublings of the well-characterized cells can then be successfully cryopreserved for decades (if not longer); and successive lots of vaccine can be created from a single tissue source. With the potential for approximately 50 population doublings from one original culture, an enormous number of cells can be grown from a single vial of early passage cells. Moreover, in comparison to primary monkey kidney cells, which have been contaminated with a wide variety of agents, including oncogenic SV40, Marburg virus, B virus, and simian cytomegalovirus, HDCS are clean.

Nevertheless, since the time of their development, many objections have been made to the use of HDCS for the production of vaccines [1, 9], including their foetal source, the hypothetical risk of cancer from cellular DNA, and the possibility that there were unknown adventitious viruses within the strains. The U.S. FDA also played a role in hampering the acceptance of HDCS cells for vaccine production.

The source of the HDCS

Although the HDCS are derived from normal cells, some individuals objected to the source, human foetal tissues, on ethical or religious grounds. Indeed, originally some of the foetuses came from Sweden, as abortion was then not legal in the USA. In addition, public health authorities believed that depending on a single cell strain like WI-38 for indispensable vaccines (i.e. «putting all your eggs in one basket») was an ill-advised policy, especially when primary tissue culture-produced vaccines were already available and considered at the time to be known, safe alternatives. With respect to this opinion, Petricciani has written that «the dogma of safe cells being limited to those derived from primary tissue had an inhibitory effect on the introduction of HDCS» [10].

Risk of cancer from the HDCS substrate

Some regulators felt that HDCS cells themselves possessed oncogenic characteristics. It was argued that HDCS cells are a multigeneration, heterozygous population; the chromosomal anomalies that developed in a small percentage of cells suggested that transformation occurs within this population of continuously cultured cells. In particular, a 1962 National Institutes of Health (NIH) committee report stated, with respect to «continuously cultured cells», that «the resultant cell populations are heterogeneous, which means that they are not precisely characterized genetically or otherwise and may be subject to random fluctuations in properties» [11, 12]. Reflecting those concerns, the present policies that govern use of HDCS cultures emphasize two points: (i) the initial seed culture must begin with a normal karyology and (ii) any chromosomal anomalies that develop as the cells age must be identified. In particular, the WHO instructs that human diploid cells should not be used «beyond two-thirds of the total number of population doublings corresponding to the average finite life of the cell», while the U.S. regulatory authorities have chosen to constrain the number of population doublings at the point that the number of chromosomal anomalies reaches an «acceptable limit» [13]. In practice, the limit is usually reached at 35 to 40 population doublings. From an experimental standpoint, it should be noted that direct injection of continuously cultured HDCS cells has not been tumorigenic in any animal model, and no tumors formed when HDCS cells were injected into human volunteers [1]. Moreover, no spontaneous transformation of HDCS cells has ever been observed [12].

Adventitious agents within the strain

The fear of transformation was bound up with concerns about human cellular DNA in the final, HDCS-derived vaccine product. At present, picograms of residual cell DNA remain in a single parenteral dose of most vaccines, and this degraded DNA could code for little [14, 15]. In addition to DNA, the possibility of unknown endogenous contaminants was raised as an argument against the use of HDCS cells for vaccine production. Using direct or indirect methods of detection, no evidence exists for an endogenous human oncogenic agent or latent virus, e.g., non-living transforming principles, slow viruses, or endogenous retrovirus, in any HDCS-derived vaccine. Furthermore, it was reasoned that, while a human source may be ideal for adapting and growing viruses which target humans, the same human source likewise eliminates the species barrier that blocks exogenous adventitious agents. Reassuringly, most contaminants, such as viruses, microbes, host proteins, and nucleic acids, can be identified and then inactivated or removed.

The role of the U.S. government

The U.S. Government's role in the history of HDCS has been complex. The 1962 rejection of the U.S. patent application, since there was no precedent at that time that the U.S. Patent Office must offer patents for living material, left the ownership of WI-38 in doubt. In 1976, the U.S. Government accused Hayflick of stealing government property, leading to a legal struggle that ended in 1981 with a complete exoneration of Hayflick and the decision that the title to WI-38 is not vested with the U.S. Government. During this nearly 20-year period of uncertain ownership, seed cultures of WI-38 were distributed freely to researchers around the world, and the WHO actively established repositories for WI-38 in the UK and the former-USSR. In the USA, the Division of Biologic Standards (DBS) initiated a project to develop non-human diploid cell lines, putting WI-38 in potential competition with these NIH, non-human cell strains. At the same time, the initially negative opinion of the FDA towards the use of HDCS cells to prepare vaccines (by maintaining the idea of a hypothetical something in the cells, such as a «human leukemia agent», that would make such vaccines risky) should not be underestimated.

Licensed Vaccines Made in HDCS [16]

At least nine human vaccines using HDCS as a substrate have been commercialized (Table 2).

Table 2: Vaccines prepared on human diploid cells.

Vaccine	Manufacturer(s)	HDCS	Purification	Adjuvants
Live				
Adenovirus	Wyeth	WI-38	–	–
Measles	Institute of Immunology, Zagreb	WI-38	–	–
Mumps	SSVI	MRC-5	–	–
OPV	Pfizer	WI-38	–	–
Rubella	Merck PMC SB SSVI	WI-38 MRC-5 WI-38, MRC-5 MRC-5	– –	– –
Varicella	Merck PMC SB	WI-38 MRC-5 MRC-5	– –	– –
Inactivated				
Hepatitis A	Merck PMC SB	MRC-5 MRC-5 MRC-5	+ + +	Alum Alum Alum
IPV	PMC	MRC-5	+	± Alum*
Rabies	PMC	MRC-5	–	–

* When combined with DTP.

Poliovirus (multiple strains)

Different strains of the three poliovirus types (type 1, type 2, type 3) were first adapted to the HDCS in 1962 [17]. Poliovirus vaccination is part of routine childhood immunization, and is also indicated for all susceptible persons. The killed virus vaccine, IPV, is injected i.m., and four doses are given during the first two years of life. For the live, attenuated vaccine, OPV, the recommended dosage schedule in the USA has been four doses, three during the first 6 to 18 months life, followed by a fourth dose at 4 to 6 years of age. OPV vaccine is given orally as drops.

Between 1991 and 1995, 6.3 million doses of an MRC-5 IPV vaccine were distributed in Canada, the U.S.A., and other countries, and there have been few reported serious side effects. (Pharmacovigilance following 697,200 doses of IPV in Canada, for example, suggests an adverse event rate of approximately 1 per 100,000 doses for either allergic reactions, fever, or hypotonic-hyporesponsive episodes, and there was a single case of thrombocytopenia reported.) For OPV, however, cases of vaccine-associated paralytic poliomyelitis occur at a frequency of about 2.5 cases per million doses, although the incidence is about 10 times greater after the first dose. The mechanism of paralysis rests on the reversion of the attenuated strains during replication in the human intestine, which occurs regardless of the cell substrate employed.

Rabies virus (Pitman-Moore L503 3M strain)

The Pitman-Moore strain of rabies virus was first grown in HDCS in 1961 [2].

Prophylactic rabies immunization is indicated for professions that involve risk of animal bite, certain laboratory workers, children living in areas with endemic rabies, or persons moving to these areas. Three doses are given over one month, either i.m. in the deltoid or i.d. In contrast, after a potentially rabid animal bite, the vaccine must be given promptly (in conjunction with rabies immune globulin, according to the severity of the bite) combined with vigorous washing. This is followed by four more vaccine doses over the month. If the victim has received previous pre-exposure vaccine prophylaxis, however, i.m. vaccination with two doses at days 0 and 3, without rabies immune globulin, is adequate.

Rabies vaccine causes type I hypersensitivity reactions in about 6% of vaccine recipients (i.e., IgE-mediated anaphylactic reactions, mostly manifesting as a rash but sometimes by signs such as angioedema or arthralgia). These reactions have been linked to antibodies directed against the human serum albumin used as a stabilizer during vaccine production, but which is then altered during the inactivation of virus by β-propiolactone. If the concentration of β-propiolactone/human serum albumin in the HDCV vaccine is reduced, using zonal centrifugation, the incidence of such IgE-mediated delayed hypersensitivity reactions is significantly reduced [18]. Rabies vaccination is also associated with type III hypersensitivity reactions (i.e., immune complex precipitation, presenting as serum sickness or vasculitis), but these are also seen with non-HDCS vaccines. In addition, a review of safety data from 7.2 million doses (1985 to 1994) of a licensed human diploid rabies vaccine (HDCV) reveals a rate for serious systemic reactions, mostly allergic reactions, that ranged from 1 to 7 cases per 100,000 doses [19]. There also have been case reports of neurologic disorders (e.g., Guillain-Barré syndrome, meningoradiculitis, and CNS demyelinization) temporally related to HDCV vaccination [20], but of uncertain causal relationship.

Rubella virus (RA27/3 strain)

Rubella virus, strain RA27/3, was first adapted to HDCS in 1965 [21]. Now a part of routine childhood vaccination to lessen the likelihood of congenital rubella, the vaccine is given twice to all prepubertal boys and girls: the first dose is given on or after the first birthday; and the second dose, either at four to six years of age or at 11 to 12 years of age. In addition, non-immune adult women are immunized before pregnancy. RA27/3 is the only strain in widespread use: at least 100 million doses have been given, as a monovalent vaccine or in combination with measles and mumps vaccines.

RA27/3 was isolated from the cells of an infected foetus, and then transferred to WI-38 cells. As an attenuated virus, it multiplies at the site of s.c. or i.m. injection and spreads to some degree like the wild virus. Low degrees of pharyngeal virus excretion and viraemia are known to be produced. Thus, side-reactions appear to be similar to a mild case of rubella, including joint symptoms that probably represent replication of the vaccine virus in synovial tissue. The arthralgia and arthritis are evidently influenced by hormonal factors, being uncommon in children but frequent in post-pubertal women. Acute arthropathy occurs in 14% of seronegative adult women; the joint symptoms are usually transitory, presumably disappearing when the immune response eliminates the virus. Rare cases of prolonged symptoms have been reported. The Institute of Medicine (USA) concluded in 1991 [22] that there is «evidence consistent with a causal relation between the currently used vaccine strain (RA27/3) and chronic arthritis in adult women». However, subsequent studies have failed to confirm that statement. Thrombocytopenia is seen after natural disease, and there have been case reports of thrombocytopenic purpura in vaccinees, although the Institute of Medicine concluded that the evidence was insufficient to show a causal relationship [22].

Measles virus (Edmonston-Zagreb strain)

The Edmonston-Zagreb strain of measles virus was first grown in HDCS in 1972 [5].

In developed countries, measles vaccine should be administered s.c. to children in two doses, the first dose given on or after the first birthday, and the second given either at four to six years or at 11 to 12 years of age.

The public health goal of measles vaccination is the identification and immunization of all susceptible person, especially pre-school-aged children. Among African infants (in Senegal, The Gambia, and Guinea-Bissau), a series of clinical trials tested the efficacy of a single injection of high titre measles vaccine. Follow-up suggested improved survival during the one to two years after vaccination among the girls who had received a standard titre Schwarz measles vaccine compared to the girls who had received either the high titre Edmonston-Zagreb or high titre Schwarz strain measles vaccines [23]. It is important to note that the vaccines were given at six months of age, much earlier than usual. This temporary difference in mortality, which did not persist among girls followed longer that one to two years after vaccination, was seen after high titre Schwarz vaccine grown in chick embryo fibroblast cultures was used, as well as the human diploid cell-adapted, high titre Edmonston-Zagreb vaccine, which exonerates the HDCS substrate from playing any role in this syndrome [24]. CNS conditions (i.e., encephalitis and encephalopathy) after vaccination have occurred in less than one instance per 2.5 million vaccine doses (1979-1986, USA). Thrombocytopenic purpura following vaccination is a rarely seen adverse event, but has been associated with the measles vaccine.

Varicella-zoster virus (Oka strain)

The Oka strain of varicella-zoster virus (VZV) was developed on HDCS as a vaccine in 1974 [25]. VZV vaccine is now indicated for the vaccination of certain susceptible immunocompromised persons and, in the USA, Japan, and Korea, for routine childhood immunizations. Children in the USA receive the VZV vaccine s.c. between 12 and 18 months of age. Unvaccinated children who have an uncertain history of chickenpox should be immunized at the 11- to 12-year-old visit.

Approximately 2 million doses have been used already in the USA (August 1995 to February 1996). The side effects of the vaccine are essentially mild varicella, including rash. With the possible exception of hepatitis, more serious systemic involvement has not been seen. Children suffering from a state of immunodeficiency–such as leukaemic children–risk the development of a varicella-like illness following administration of this live, attenuated virus. The fact that the vaccine is a relatively crude suspension of virus containing much HDCS tissue is particularly relevant evidence for the safety of HDCS.

Mumps virus (Rubini strain)

The Rubini strain of mumps virus, first adapted to the HDCS in 1986 [26], is licensed as a vaccine in Switzerland and in Italy, and is distributed to some other countries. Mumps immunization is part of the standard childhood schedule, but is also important for any remaining susceptibles such as children who are approaching puberty and international travellers. Mumps vaccine should be administered in two doses, the first given on or after the first birthday, and the second given at four to six years of age or at 11 to 12 years of age.

Mumps vaccine-associated meningitis is not associated with the HDCS-grown strains; it has been described after vaccination with the chicken or quail embryo fibroblast mumps vaccine strains, such as Urabe, Leningrad, Hoshino, and Torii.

Hepatitis A virus (HM-175, CR-326-F', and GBM strains)

Hepatitis A virus strain HM-175 was first adapted to HDCS in 1988 [27].

The currently licensed vaccines are indicated for high-risk adults, which includes travellers to endemic areas, military personnel and food handlers; routine childhood HAV vaccination in endemic areas seems likely. The vaccine is injected i.m. into the deltoid. The hepatitis A vaccines are administered by a two-dose schedule; the paediatric dose is half of the adult dose.

From post-marketing surveillance following administration of approximately 20 million doses of one of the licensed hepatitis A vaccines, there have been isolated reports of major side-effects, in particular, polyradiculoneuritis and (spontaneously resolving) leukocytoclastic vasculitis [19]. Leukocytoclastic vasculitis has been reported after hepatitis A itself.

Lessons Learned

The development of HDCS-derived viral vaccines has been a challenge, both for the researchers and physicians who created these vaccines and for the government agencies charged with regulating them. The control of vaccines manufactured on this substrate has provided lessons about (i) laboratory screening for adventitious agents and contaminants during the production of vaccines, (ii) pharmacovigilance of a vaccine during its development and public use, and (iii) reasonable ways to approach regulatory issues.

Concerns over safety preceding vaccine licensing impelled manufacturers and regulators to develop more elaborate, sensitive, and sophisticated screening methods for viruses, microbes, host proteins, and nucleic acids. In applying such standards to potential virus contamination, for example, a 1986 WHO Study Group on Biologicals classified human diploid cells as being of low risk for carrying pathogenic viral agents [28].

At the same time, networks of pharmacovigilance were established to monitor vaccine-associated side effects and to detect rare, serious events, by following the generally accepted concept that «the safety of the final product is the appropriate endpoint in deciding on the acceptability of cell substrates [28].» Table 3 summarizes the serious events that have been associated with HDCS viral vaccines. These rare reactions appear to reflect the activity of the live viruses replicated in the HDCS or of the inactivated viruses or other proteins added during manufacture. There is no credible association of reactions to the HDCS substrate or a hypothetical contaminant derived from it.

Widely-used, HDCS-derived human vaccines against rabies, rubella, and polio (among others) now have at least a 20 year record of vaccine safety after extensive, ongoing pharmacovigilance «without any evidence of any adverse reaction attributable to the substrate, or any report of the detection of a virus contaminant, either infectious, latent, or oncogenic». This led the same author, David Magrath, to conclude that «the two well characterized cell lines WI-38 and MRC-5 may perhaps be regarded as truly safe substrates [14]». Moreover, some vaccines, like hepatitis A and varicella, could not easily be made in other substrates. Thus, HDCS have been a boon to vaccinology.

Finally, looking toward the future,

Licensing authorities have a responsibility to define outstanding scientific issues on which answers are required in order for them to make decisions. It does not serve the interest of health in any part of the world for regulatory agencies to delay decisions by failing to ask specific and relevant scientific questions that can be answered [10].

Table 3: Rare, serious reactions attributed to vaccines made in HDCS.

HDCS vaccine	**Serious reactions due to vaccination**
IPV	Thrombocytopenia
OPV	Vaccine associated paralytic poliomyelitis*
Rabies	IgE mediated anaphylactic reactions*, particularly after booster doses; type III hypersensitivity reactions (i.e., immune complex precipitation, presenting as serum sickness or vasculitis); and case reports of neurologic disorders (e.g., Guillain-Barré syndrome, meningoradiculitis, or CNS demyelinization)
Rubella	Acute arthropathy* and chronic arthritis, particularly in adult women; thrombocytopenic purpura*
Measles	Thrombocytopenic purpura; encephalitis and encephalopathy; immune depression after high dose vaccine*
Varicella-zoster	Varicella-like illness*
Mumps	Aseptic meningitis*
Hepatitis A	Polyradiculoneuritis and (spontaneously resolving) leukocytoclastic vasculitis

* Reactions for which there is good evidence for a causal relation.

Acknowledgements

The authors appreciate the thoughtful comments of Drs. L. Hayflick, B. Montagnon, and J. Shindman during the preparation of this manuscript.

References

1 Hayflick L, Plotkin SA, Stevenson RE: History of the acceptance of human diploid cell strains as substrates for human virus vaccine manufacture, in Petricciani JC, Hennessen W (eds): Cells, Products, Safety. Dev Biol Stand. Basel, Karger, 1987, vol 68, pp 9-17.

2 Hayflick L, Moorhead PS: The serial cultivation of human diploid cell strains. Exp Cell Res 1961;25:585-621.

3 Chanock RM, Ludwig W, Heubner RJ, Cate TC, Laing-Wei C: Immunization by selective infection with type 4 adenovirus grown in human diploid tissue culture. I. Safety and lack of oncogenicity and tests for potency in volunteers. J Am Med Assoc 1966;195:453-459.

4 Ikic D: Recent information on poliomyelitis vaccine, live, oral, prepared in human diploid cell strain system. Progr Immunobiol 1964;2:305-310.

5 Andzhaparidze OG, Dosser EM, Rapoport RI, Unanov SS, Dorofeev VM, Iablokova ML, Sokolova TM, Avakova AN, Svezhinina IA, Froltsova AE, et al: Live measles vaccine from human diploid cells (safety, reactogenicity and immunogenicity). Vopr Virusol 1968;87(2):411-418.

6 Plotkin SA, Cornfeld D, Ingalls TH: Studies of immunization with living rubella virus: trials in children with a strain cultured from an aborted fetus. Am J Dis Child 1965;110:381-389.

7 Hayflick L: History of the development of human diploid strains, in Characterization and Uses of Human Diploid Cell Strains. Dev Biol Stand. Basel, Karger, 1963, vol 8, pp 37-54.

8 Jacobs JP, Jones CM, Baille JP: Characteristics of a human diploid cell designated MRC-5. Nature 1970;227:168-70.

9 Hayflick L: History of cell substrates used for human biologicals, in Hayflick L, Hennessen W (eds): Continuous Cell Lines as Substrates for Biologicals. Dev Biol Stand. Basel, Karger, 1989, vol 70, pp 11-26.

10 Petricciani JC: Cells, science and health, in Hayflick L, Hennessen W (eds): Continuous Cell Lines as Substrates for Biologicals. Dev Biol Stand. Basel, Karger, 1989, vol 70, pp 3-10.

11 Committee Report: Continuously cultured tissue cells and viral vaccines. Science 1963;139:15-20.

12 Hayflick L, Moorhead PS, Pomerat CM, Hsu TS: Choice of a cell system for vaccine production. Science 1963;140:766-768.

13 Hayflick L: Mislabeling, contamination, and other sins of cultured flesh, in Perkins FT, Regamey RH (eds): Standardization of Cell Substrates for the Production of Virus Vaccines. Dev Biol Stand. Basel,Karger, 1977, vol 37, pp 5-15.

14 Magrath DI: Safety of vaccines produced in continuous cell lines, in Horaud F, Brown F (eds): Virological Aspects of the Safety of Biological Products. Dev Biol Stand. Basel, Karger, 1991, vol 75, pp 17-20.

15 Horaud F: Viral vaccines and residual cellular DNA. Biologicals 1995;23:225-228.

16 Plotkin SA, Mortimer EA: Vaccines. 2nd ed. Philadelphia: W.B. Saunders Company; 1994.

17 Hayflick L, Plotkin SA, Norton TN, Koprowski H: Preparation of poliovirus vaccines in a human fetal diploid cell strain. Amer J Hyg 1962;75:240-258.

18 Fishbein DB, Dreesen DW, Holmes DF, Pacer RE, Ley AB, Yager PA, Sumner JW, Reid Sanden FL, Sanderlin DW, Tong TC, Kemp DT: Human diploid cell rabies vaccine purified by zonal centrifugation: A controlled study of antibody response and side effects following primary and booster pre-exposure immunizations. Vaccine 1989;7(5):437-442.

19 Hessel L: Safety report: inactivated HDC vaccine (1985-1994). Lyon, France. Pasteur Mérieux Sérums & Vaccins. 1996;

20 Yalcin D, Mutlu A, Forta H: MRI findings in acute disseminated encephalomyelitis due to immunization with human diploid cell rabies vaccine. Med Bull Istanb Med Fac 1996;29(1):97-100.

21 Plotkin SA, Farquhar JD, Katz M, Buser F: Attenuation of RA27/3 rubella virus in WI-38 human diploid cells. Am J Dis Child 1969;118:178-85.

22 Institute of Medicine (U.S.). Committee to Review the Adverse Consequences of Pertussis and Rubella Vaccines, in Howson CP, Howe CJ, Fineberg HV (eds): Adverse effects of pertussis and rubella vaccines. Washington, D.C. National Academy Press, 1991.

23 Aaby P, Samb B, Simondon F, Knudsen K, Seck AMC, Bennett J, Markowitz L, Whittle H: Five year follow-up of morbidity and mortality among recipients of high-titre measles vaccines in Senegal. Vaccine 1996;14(3):226-229.

24 Aaby P, Samb B, Simondon F, Knudsen K, Seck AM, Bennett J, Whittle H: Divergent mortality for male and female recipients of low-titer and high-titer measles vaccines in rural Senegal. Am J Epidemiol 1993;138(9):746-755.

25 Takahashi M, Otsuka T, Okuno Y, Asano Y, Yazaki T, Isomura S: Live vaccine used to prevent the spread of varicella in children in hospital. Lancet 1974;2:1288-1290.

26 Glück R, Hoskins JM, Wegmann A, Just M, Germanier R: Rubini, a new live attenuated mumps vaccine virus strain for human diploid cells, in Van Wezel AL, Hennessen W (eds): Use and Standardization of Combined Vaccines. Dev Biol Stand. Basel, Karger, 1986;65:29-35.

27 Peetermans J: Production, quality control and characterization of an inactivated hepatitis A vaccine. Vaccine 1992;10(Suppl 1):99-101.

28 Petricciani JC: Regulatory philosophy and acceptability of cells for the production of biologicals, in Horaud F, Brown F (eds): Virological Aspects of the Safety of Biological Products. Dev Biol Stand. Basel, Karger, 1991, vol 75, pp 9-15.

Dr. S.A. Plotkin, Pasteur Mérieux Connaught, 3, avenue Pasteur, B.P. 10, F-92430 Marnes la Coquette, France.

Brown F, Griffiths E, Horaud F, Petricciani JC (eds): Safety of Biological Products Prepared from Mammalian Cell Culture.
Dev Biol Stand. Basel, Karger, 1998, vol 93, pp 109-117

Chromosomal Characterization of MRC-5 Cell Banks Utilizing G-Banding Technique

M. Rosolowsky, R. McKee, W. Nichols, B. Garfinkle

Merck Manufacturing Division and Merck Research Laboratories,
Merck & Co., West Point PA, U.S.A.

Key words: MRC-5 cells, karyology, G-banding, 7;12 translocation, cytogenetic criteria.

Abstract: We have performed chromosomal monitoring on 18 MRC-5 cell banks using the G-banding technique. A higher frequency of structural abnormalities and hyperdiploidy was detected with respect to the numerical values that were established with conventional staining techniques and reported in the 1979 Ad Hoc Committee guideline on karyology controls of human cell substrates. These numerical criteria have been adopted by international regulatory agencies to release human diploid cell banks (WI-38 and MRC-5) for vaccine manufacture. In the process of characterizing these 18 cell banks, a 7;12 translocation clone was detected in 7 cell banks at a frequency ranging from 0.2 to 5.6%. The presence of the t(7;12) appears to be biphasic. At low population doubling levels (PDLs) (< 30), t(7;12) is rarely observed. However, the incidence of t(7;12) increases and plateaus between PDL 40-50. A decrease in frequency is observed at higher PDLs. Before senescence of the cell bank, t(7;12) is not observed. Investigation of the growth characteristics of MRC-5 cells revealed that cell banks containing the translocation senesced at similar PDLs compared to MRC-5 cells with no detectable 7;12 translocation. In addition, MRC-5 cell banks containing the t(7;12) have successfully completed tumorigenicity testing in a nude mouse model. We conclude that MRC-5 cells obtained from either National Institute for Biological Standards and Control (NIBSC) or American Type Culture Collection (ATCC) contain a 7;12 translocation at a low frequency. This abnormality does not provide MRC-5 cell bank mass cultures with a growth advantage nor is it tumorigenic in nude mice. Furthermore, the presence of this clone and employment of the G-banding technique may be responsible for the increased observation of structural abnormalities in our laboratories. In view of this information, the cytogenetic criteria that were established in 1979 with conventional staining techniques are not appropriate for human diploid cell banks that are examined with more sensitive methodology. Since it has been recognized that MRC-5 and WI-38 cells are safe biological substrates, we recommend that MRC-5 and WI-38 cell banks should only be identified by using an appropriate identity test and should not require any chromosomal analysis before being used as a cell substrate for the manufacture of live virus vaccines.

INTRODUCTION

Over the last 30 years, human diploid cells (MRC-5 or WI-38) have been successfully used as cell substrates to manufacture vaccines against measles, mumps, rubella, polio, rabies, varicella and hepatitis A [1]. Before vaccine production, manufacturers are required to ensure that cell substrates are free of adventitious agents and any neoplastic properties. Chromosomal characterization (karyology), in vitro life span and injection of immunocompromised mice with cells (tumorigenicity) are three procedures that have been used to demonstrate that cell substrates lack neoplastic properties. These standard tests were established in the 1960s to verify the safety of vaccines made from human cell substrates.

MRC-5 cells were isolated from lung tissue of an aborted male foetus and were characterized according to internationally recommended criteria such as virus susceptibility, cell growth characteristics, absence of adventitious agents and neoplastic properties, and karyology [2]. International meetings were held in Geneva in 1976 and in Lake Placid in 1978 to discuss karyology guidelines for human cell substrates. A report was published with karyological control recommendations based upon a statistical analysis of conventionally stained WI-38 cells [3]. These recommendations were subsequently adopted by international regulatory agencies as requirements for human cell substrates, including MRC-5 and WI-38 cells, used for the production of biological products. These karyological criteria are still applied today.

Although it was acknowledged in the Lake Placid report that the «incidence of karyotypic abnormalities... that are detectable with the greater resolution provided by banding should be evaluated upon accumulation of a larger data base», no such karyological analysis of a large data base of banded human diploid cells has been published. The purpose of this paper is to report on the karyology of 18 MRC-5 cell banks that have been examined in our laboratories by the Giemsa-banding technique. The incidence of abnormalities observed with this higher resolution technique was compared to the incidence found previously for conventionally stained human diploid cells. During this study, a 7;12 translocation clone was identified in MRC-5 cells, and its benign properties were characterized and reported in this paper. Based on our findings, modifications of the karyological requirements are recommended for MRC-5 and WI-38 cells.

Materials and Methods

Cell culture and metaphase preparation

Low population doubling level MRC-5 cells (≈ PDL 8) were obtained from either ATCC or NIBSC. Cells were expanded in either serum-supplemented EBME or Williams E media to prepare cell banks. Before use, MRC-5 cells and tissue culture reagents were tested for the absence of mycoplasma, microbial, fungal and viral contaminants.

To prepare cells for chromosomal monitoring, cells were accumulated in metaphase by addition of colcemid. Cells were then washed and harvested with a trypsin solution followed by suspension in hypotonic KCl. Cells were fixed in methanol/glacial acetic acid, washed and diluted to the proper density, and dropped on to glass microscope slides to prepare metaphases. To band chromosomes, the slides were immersed in a trypsin solution. Slides were washed before being stained in 10% Giemsa stain solution.

Chromosomal analysis

Metaphases were selected for analysis manually or by utilizing a Cytoscan. At least 500 metaphases were analyzed under oil immersion optics (1000X magnification) for exact counts and chromosomal abnormalities. The chromosome complement including the number of chromosomes, the sex chromosomes and any abnormalities were scored based on the International System for Human Cytogenetic Nomenclature (ISCN 1985).

RESULTS/DISCUSSION

Karyological analysis of MRC-5 cells by G-banding

Karyological analysis of 18 MRC-5 cell banks was performed using the Giemsa staining/banding (G-banding) technique. A total of 8579 metaphases were analyzed for cells harvested at population doubling levels ranging from 30 to 48, with the majority of cell banks between PDL 35-40. In most cases, 500 metaphases were scored per cell bank. The results are summarized in Table 1. For comparison, the 95% upper confidence limits as recommended by the 1979 Ad Hoc Committee [3] are listed. It is apparent that the 95% upper confidence limits of hyperdiploidy and structural abnormalities are significantly greater in our analysis than what was reported in the 1979 Ad Hoc Committee paper. It is likely that the differences observed in structural abnormalities are due to an alternative methodology. Examination of cells at higher population doubling levels, differences between analysts in slide preparation, or a combination of these factors could influence the incidence of aneuploidy and structural abnormalities.

Historically, chromosomal analysis of MRC-5 cells and WI-38 cells had been performed with conventional staining methods. While it is easier to identify breaks and gaps with conventional staining, it is more difficult to determine structural rearrangements. Since we examined metaphases with G-banding, it is easier to detect stable structural rearrangements, and this could have accounted for the higher incidence of structural abnormalities observed. In our study, cells were examined using G-banding at an average PDL of 36. The majority of the data used to compile the 1979 Ad Hoc report were from WI-38 cells at PDL 27-33. It has been shown that aneuploidy or structural abnormalities may increase when human diploid fibroblasts are analyzed later in their lifespan [4, 5]. Examination of MRC-5 cells at higher PDLs could have contributed to the higher incidence of abnormalities in our study. It has been reported in the 1979 Ad Hoc report that structural abnormalities were not homogeneous among laboratories. This suggests that even when cells are examined by the same methods (i.e. conventional staining), there are significant differ-

Table 1: Karyological analysis of MRC-5 manufacturer's working cell banks.

Parameter	Average incidence (%)[1]	95% upper limit[1]	95% upper limit[2]
Hypodiploid	12.1	21.2	18
Hyperdiploid	0.8	2.6	1.0
Polyploid	1.9	3.7	3.4
Breaks	0.9	2.5	5.2
Structural abnormalities	1.4	3.8	2.0

1 Current study which included analysis of 18 MRC-5 MWCBs;
Total number of metaphases examined = 8579;
Typically 500 metaphases scored per MWCB.

2 Ad Hoc Committee Report on karyological controls of human cell subtrates. J Biol Stand 1979;7:397-404.

ences between laboratories. Similarly, analyst differences in scoring metaphases for structural abnormalities may have contributed to the higher incidence of abnormalities in this study.

Observation of t(7;12)

During the course of these analyses, a 7;12 translocation was observed at varying frequencies (Fig. 1). The t(7;12) was observed in seven of the 18 MRC-5 cell banks studied. The presence of t(7;12) would also contribute to the elevated incidence of structural abnormalities observed in MRC-5 cells. If the incidence of t(7;12) was eliminated from the calculation of the 18 MRC-5 MWCBs examined (Table 1), the average incidence of structural abnormalities would decrease to 1.0% with a 95% upper limit of 2.8%.

The appearance of clonal events in human diploid cell fibroblasts is not unusual, and it has been demonstrated that identification frequently requires use of banding

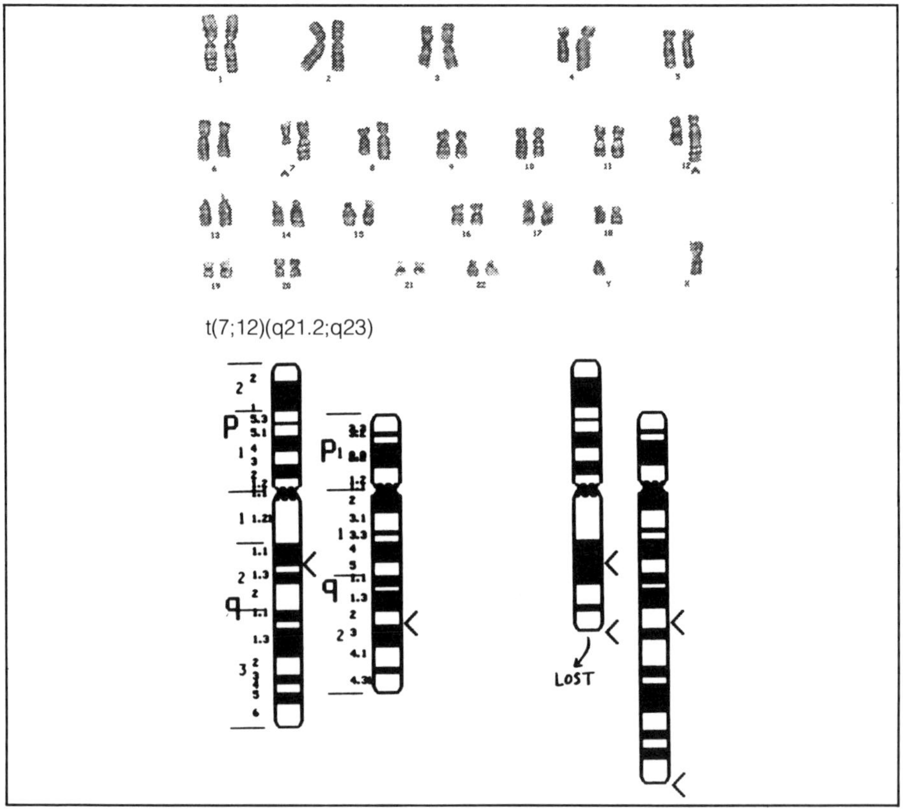

Fig. 1: Karyotype of MRC-5 MWCB containing 7;12 translocation. Inset shows ideogram of t(7;12) demonstrating break points.

methodologies. Chen and Ruddle [6] identified recurring chromosomal rearrangements in WI-38 cells. Three chromosomal modifications were detected when banding techniques were employed: a) deletion of the short arm of chromosome 1 (1p-); b) deletion of the short arm of chromosome 4 (4p-); and c) a translocation of the distal two-third segment of 11q to the short arm of chromosome 17 t(11q-,17). Interestingly, if conventional staining methods were used the t(11q-,17) would appear to be a pseudo-diploid (+C/-E). Likewise, conventional staining methodologies have been used almost exclusively to perform chromosomal analysis of MRC-5 cells. The t(7;12) translocation could probably be detected with a very careful karyotype of unbanded chromosomes as a chromosomal change, but the specific change could not be interpreted. Because of the altered size and centromere placement of the translocation this would probably appear to be an extra B group chromosome (chromosome 4, 5) in an unbanded preparation with loss of a C group chromosome (+B/-2C/+E). This would make the metaphase appear to be pseudo-diploid. Clonal translocations and other clonal structural chromosomal abnormalities have been observed frequently in human diploid fibroblasts from normal adults when banding techniques are used [7, 8].

Incidence of t(7;12) over lifespan and senescence

Chromosomal monitoring was performed on several cell banks to determine the incidence of t(7;12) over the lifespan of each cell bank. Figure 2 shows data of three cell banks which had multiple analyses for t(7;12) during the course of continuous culture. The 7;12 translocation is not observed at low PDLs. However, the incidence of t(7;12) increases and seems to plateau between PDLs 40 and 50. A decrease in frequency of t(7;12) is seen at higher PDLs (Figs. 2A and B). This profile suggests that the 7;12 clone has a limited, enhanced growth rate, but it then eventually senesces. Different interdivisional times of clonal daughter cells have long been recognized using time-lapse cinematography. A clonal chromosomal alteration permits recognition of different divisional times in mass culture populations [9]. A biphasic profile of clonal events is similar to what has been reported by Kano and Little who showed clonal expansion and attenuation in X-irradiated human diploid fibroblasts [10]. Cells from their study did not lose growth control because overall senescence was observed at the expected population doubling levels. For one cell bank in our analysis, the t(7;12) was never observed at any point throughout its lifespan (Fig. 2C). This indicates that the 7;12 translocation is a random, low frequency event. The fact that it was not seen in one cell bank throughout its lifespan (Fig. 2C) argues against the translocation being an artifact of slide preparation and sample processing.

Table 2 shows the PDL at senescence for cell banks along with the incidence of t(7;12) that had been observed over the course of several cell expansions for each bank. Cells senesced between 62 and 73 population doubling levels, demonstrating that there was no overall growth advantage associated with the presence of the 7;12 translocation. This finding is consistent with the lifespan study described above. Since the completion of this study, cell culture conditions have been optimized to achieve senescence at higher PDLs (80-90). As before, cell banks containing a t(7;12) senesce at a similar PDL as those cell banks in which the t(7;12) is not observed. The fact that MRC-5 cells senesce even when exhibiting the t(7;12) is

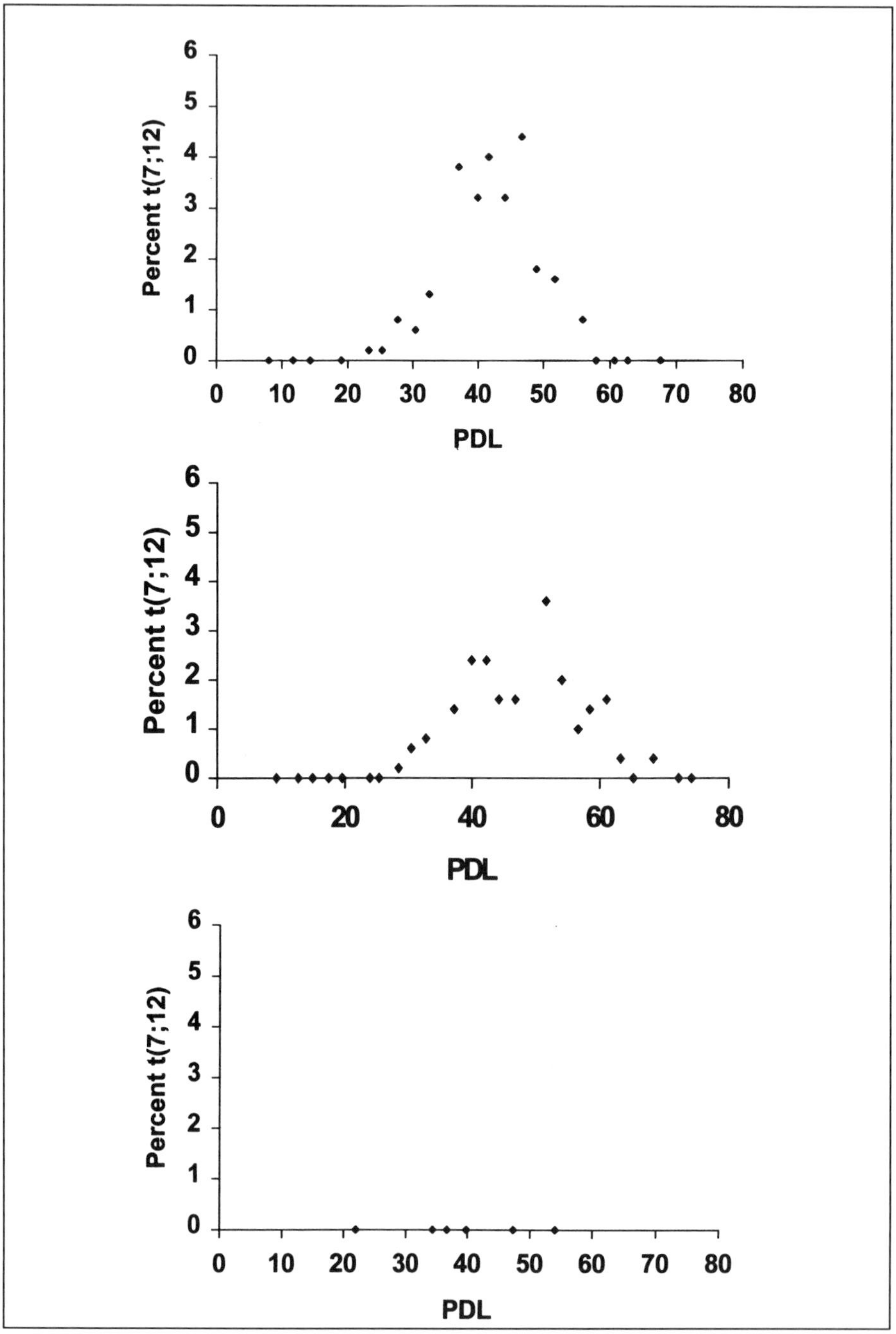

Fig. 2: Graphs of the incidence of t(7;12) in MRC-5 cells derived from three different Jacob's pool ampoules (A,B, and C) during the lifespan of cells in continuous culture.

Table 2: Senescence of MRC-5 cells and incidence of t(7;12).

Cell bank	% 7;12 translocation observed during multiple cell expansions (PDL)	PDL at senescence
1	0.2% (18.0)	
	5.6% (30.8)	
	0.4% (33.1)	
	0% (33.8)	
	2.6% (37.4)	
	0% (46.1)	64.8
2	0.4% (18.3)	
	2.6% (33.3)	
	2.5% (36.4)	
	0.4% (43.9)	
	0% (45.9)	
	1.4% (56.2)	73.2
3	3.2% (32.1)	
	0% (39.6)	61.9
4	0% (31.3)	
	0% (36.7)	62.0
5	0% (21.9)	
	0% (34.4)	
	0% (36.7)	
	0% (39.8)	
	0% (47.3)	
	0% (54.0)	67.7

conclusive evidence that the cells have not been transformed as a result of this structural abnormality.

Tumorigenicity

Cells from each of the 18 cell banks have successfully completed tumorigenicity studies. In these studies, each of ten nude mice (nu/nu) was injected with one or 10 million MRC-5 cells and compared to mice receiving HeLa cells (positive control) or culture media (negative control). No histological evidence of tumour formation was found in any mouse treated with MRC-5 cells after three or six weeks. In contrast, mice injected with HeLa cells developed palpable nodules and progressing tumours.

To assess the tumorigenic potential of MRC-5 cell banks in which the t(7;12) was observed more carefully, a special study was designed. MRC-5 cell banks with 0, 2.5, 3.0, 4.0 or 5.6% t(7;12) were injected into mice at a concentration of one million or 10 million cells/mouse. To determine the sensitivity of the nude mice test system, HeLa cells were injected into mice at 10-fold increments from 10^1 to 10^6 HeLa cells mixed with MRC-5 cells to achieve a dose of one million total cells/mouse. A subset of mice in each group was necropsied at days 21 and 42 post injection to monitor the histological progression of tumour formation. No histopathological evidence of tumours was detected in any of the over 200 mice injected with MRC-5 cells alone. In contrast, there was a clear concentration and time-related increasing incidence and size of HeLa cell masses. Histomorphologically, the HeLa cell masses consisted of sheets of large, pleomorphic, continually growing anaplastic cells with prominent nucleoli. As few as ten HeLa cells injected into mice resulted in palpable, neoplastic masses in nude mice. This study indicated that our nude mouse model is able to detect tumours when the concentration of HeLa cells is as low as 0.001% (10 HeLa cells in 999,990 MRC-5 cells), and that mice injected with MRC-5 cells containing up to 5.6% t(7;12) did not show evidence of tumours. This latter observation is consistent with the wealth of knowledge on the properties of MRC-5 cells. In fact, the European Pharmacopoeia has recognized the safety of WI-38 and MRC-5 cells by not requiring tumorigenicity testing of these cell substrates [11].

Summary/Conclusion

Currently, vaccine manufacturers must comply with the numerical criteria for karyological parameters as established in the 1979 Ad Hoc Committee report in order to use human diploid cells as cell substrates for biologicals. The data provided in our report support the notion that populations of cells which do not meet these criteria when examined using newer technologies (G-banding) can still be considered normal and safe for vaccine manufacture. Clonal chromosomal events, such as t(7;12), were not uniformly lethal, nor did they confer tumorigenic properties to the cell. Our study shows that numerical karyology criteria may be misleading in the case of clones which have frequencies that increase or decrease at various PDLs. Consequently, adherence to an anachronistic karyology requirement results in the unnecessary discard of a limited and valuable resource of WI-38 and MRC-5 cells. Our data verify what the authors of the 1979 Ad Hoc Committee paper expected, that employment of the high resolution G-banding technique may result in detection of a higher incidence of structural abnormalities.

The observation that MRC-5 cells contain a t(7;12), senesce and are non-tumorigenic raises questions about the value of requiring these cells to pass karyological numerical criteria which were established by conventional staining techniques. Consequently, application of the 1979 Ad Hoc Committee numerical criteria to release human diploid cells is archaic and inappropriate. We recommend that MRC-5 and WI-38 cell banks should only be identified as human diploid cells and should not require any chromosomal analysis before being used as cell substrates for the manufacture of live virus vaccines.

REFERENCES

1 Hayflick L, Plotkin S, Stevenson RE: History of the acceptance of human diploid cell strains as substrates for human virus vaccine manufacture, in Petricciani JC, Hennessen W (eds): Cell, Products, Safety. Dev Biol Stand. Basel, Karger, 1987, vol 68, pp 9-17.

2 Jacobs JP, Jones CM, Baille JP: Characteristics of a human diploid cell designated MRC-5. Nature 1970;227:168-170.

3 Committee report. Report of ad hoc committee on karyological controls of human cell substrates. J Biol Stand 1979;7:397-404.

4 Rubin BA, Minecci LC, Tint H: Karyologic characteristics of the WI-38 diploid cell system-general comments. Nat Cancer Inst Mono 1968;29:97-106.

5 Hayfick L: The limited in vitro lifetime of human diploid cell strains. Exp Cell Res 1965;37:614-636.

6 Chen, TR, Ruddle FH: Chromosome changes revealed by the Q-banding method during cell senescence of WI-38. Proc Soc Exp Biol Med 1974;147:533-536.

7 Harnden DG, Benn PA, Oxford JM, Taylor AMR, Webb TP: Cytogenetically marked clones in human fibroblasts cultured from normal subjects. Somatic Cell Genetics 1976;2:55-62.

8 Littlefield LG, Mailhes JB: Observations of de novo clones of cytogenetically aberrant cells in primary fibroblast cell strains from phenotypically normal women. Am J Hum Genet 1975;27:190-197.

9 Absher PM, Absher RG, Barnes WD: Genealogies of clones of diploid fibroblasts. Exp Cell Res 1974;88:95-104.

10 Kano Y, Little JB: Mechanisms of human cell neoplastic transformation: X-ray-induced abnormal clone formation in long-term cultures of human diploid fibroblasts. 1985.

11 Regulatory monograph: Vaccina ad usum humanum: human diploid cells for production of vaccines. European Pharmacopoeia 1994;153-5-153-6.

Dr. M. Rosolowsky, Biological Product Release, Merck & Co., Inc., P.O. Box WP 36A-211, West Point, PA 19486-0004, USA.

Brown F, Griffiths E, Horaud F, Petricciani JC (eds): Safety of Biological Products Prepared from Mammalian Cell Culture.
Dev Biol Stand. Basel, Karger, 1998, vol 93, pp 119-123

Experience with the Vero Cell Line

B.J. Montagnon, J.C. Vincent-Falquet

Pasteur Mérieux Sérums & Vaccins, Marcy L'Etoile, France

Abstract: The Vero cell line has been managed with the Cell Bank system to produce at the 142nd passage IPV, OPV and rabies vaccines since 1982 by Pasteur Mérieux Sérums & Vaccins (PMsv). The safety of the cell line was regularly validated at the Working Cell Bank (WCB) level according to the WHO and European Pharmacopoeia requirements for absence of bacteria, fungi, mycoplasma and viruses. A special emphasis was devoted to research on the absence of simian viruses (SV40, SIV, Retro-D virus and simian CMV). All these specific researches were negative.
At a low level of passage, the Vero cells are not tumorigenic.Vaccines have been prepared in low passage level Vero cells and together with the excellent downstream purification have resulted in excellent safety as attested by pharmacovigilance of more than 100 million doses of IPV during 12 years, more than 20 million doses of rabies vaccine during 10 years and more than 1 billion of OPV during eight years.

History of the Vero Cell Line

The Vero cell line was initiated in 1962 [1] and has a well recorded history, with its description updated in English in 1987 [2]. Later, transmission was made to the NIH in the U.S. in 1964, at the 93rd passage level. It was transferred to the ATCC in 1969 at the 113th passage level and some additional passages were used to secure a cell line free of mycoplasma to be stored at the repository level of passage [3, 4]. At PMsv, one ampoule of passage 124 was received from ATCC in 1979 and Primary Cell Banks were prepared [5]. Secondarily, four WCBs have been extensively used after testing for validation. Three of them are constantly in use for production of the three vaccines mentioned: IPV, rabies and OPV. Details of these developments were published regularly [6-8].

Tumorigenicity

The absence of tumorigenicity of the PMsv Vero Cell Line is regularly tested for all the WCBs by an in vivo method in immunosuppressed newborn rats [9] and the results are satisfactory to the 169th passage level. From passage 191, progressive nodules appeared which are tumorigenic with presence of metastases as observed with the tumorigenic HeLa cell line (Table 1). DNA extracted from the Vero cells, checked by the in vitro NIH test of cellular NIH 3T3 transformation, was regularly negative [10].

Table 1: Tumorigenicity of PMsv Vero Cell Line.
a) Cumulative results for Vero cells between passage 137 to 159;

Cell strains	Passages	Sub-cutaneous tumoral nodules		Lung metastases	
		Frequency	%	Frequency	%
Vero*	137 to 159	0/225	0	0/225	0
Vero°	146	0/10	0	0/10	0
Hela*	195 to 253	236/236	100	139/236	58,9
Hela°	235	10/10	100	4/10	40
HEP 2	60 to 133	111/114	97	62/114	54,3
K B	26 to 70	22/40	55	18/40	45

* 10^6 cells per rat.
° 10^7 cells per rat.

b) Results for PMsv Vero cells according to different passage levels.

Cell strains	Passages	Number of cells per rat	Sub-cutaneous tumoral nodules			Lung and node metastases
			Average diameter (mm)		Historical examination	
			d + 7	d + 21	at d + 21	Frequnecy
Vero	142	10^6 10^7	0.49 0.53	0 trace	0.10 10/10*	0/10 0/10
Vero	161	10^7	0.79	trace	2/10*	0/10
Vero	169	10^6 10^7	0.43 0.72	0 trace	1/8* 6/9*	0/8 0/9
Vero	191	10^6 10^7	0.63 0.89	0.77 1.01	10/10° 9/9°	4/10 2/9
Vero	211	10^6 10^7	0.72 1.13	1.12 1.67	6/6° 10/10°	4/6 5/10

* regressing nodules.
° progressing nodules.

Specific Virus Research

Tests for absence of viral contamination are performed regularly at the WCB level, and at least 10 PDL beyond the production level, according to the WHO-Technical report N° 745 (1987) and again for each production cell batch, observing the absence of haemadsorbing and non-haemadsorbing viruses by inoculation of control cell cultures and subcultures of the supernatants. On the other hand, some specific viruses were verified for their absence in some specialized sub-contracting laboratories.

(i) HD tumorvirus DNA sequence by G.Sauer according to published study [11];

(ii) SIVAGM or HIV-1 by Dr. Ph.Kanki, in the laboratory of Prof. M.Essex, at Harvard School of Medicine, by culture assays and cell DNA sequence analysis;

(iii) Mason-Pfizer (retro-D) virus by E.Hunter, at Birmingham (Alabama) by cell DNA sequence analysis;

(iv) Simian CMV by A.Bailey from Q.One Biotech. (Glasgow) by cell DNA sequence analysis.

All these researches confirmed the absence of these viruses or sequence or partial sequence integration in the cell DNA sequence of the PMsv Vero cell line.

Purification of Vaccine

The downstream purification applied to the harvested virus suspension decreases the cellular DNA to a very low level for OPV and IPV vaccines [12]. A series of results for IPV production is shown in Figure 1.

Conclusion

The PMsv Vero cell line is also indirectly validated by the WHO-collaborative study performed on the WHO-Vero cell bank. Especially by Southern-blot technique, no viral sequence was observed for common simian viruses known to have an oncogenic potential: herpesvirus B, cytomegalovirus, adenovirus, papillomavirus, nor for SIV. All these results were reported by F. Horaud [13].

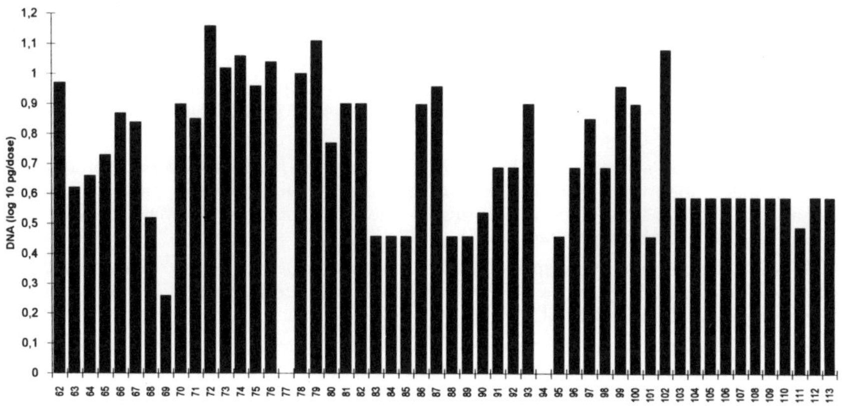

Fig. 1: Residual cellular DNA in several consecutive lots.

Using the well defined Vero cell line as cell substrate for preparing IPV, OPV and rabies vaccines, PMsv received excellent reports of safety during clinical studies. Since 1983, more than 100 million doses of IPV have been applied throughout the world, either in combined vaccines, or as IPV alone in USA, Canada and Australia.

The pharmacovigilance reporting was known for at least 78 million doses with only rare events. For rabies vaccine, licensed in France since 1985 and progressively in 50 other countries, all the reports of efficacy and safety are in favour of an excellent «tolerance». Since 1989, more than one billion doses of OPV have been applied, especially through UNICEF and PAHO tenders participating in the elimination of the wild polio strains from the Central and South America. The way is now open for the eradication of polio paralysis as a realistic goal and the use of Vero cell line as cell substrate will be a positive contribution.

Acknowledgements

We thank all the numerous colleagues involved in such work and more especially Dr J.Cl. Moulin, Cl. Goldman, B. Fanget, P. Fournier, H. Chalumeau, J.F. Saluzzo, F. Migayron, R. Vinas, R. Bandet and Dr S. Plotkin for their kind review of the manuscript.

References

1 Yasumura Y, Kawakita Y: Studies on SV40 in tissue culture. Preliminary step for cancer research in vitro. Nihon Rinsho (in Japanese) 1963;21:1201-1215.

2 Yasumura Y, Kawakita Y: Studies on SV40 in tissue culture. Preliminary step for cancer research in vitro. Nihon Rinsho (in Japanese) 1963;21:1201-1215. English translation, in Simizu B, Terasima T (eds): Vero Cells, Chiba University, Japan, 1987, pp 2-19.

3 Earley E, Johnson M and KM: The lineage of the Vero, Vero 76 and its clone C1008 in the United States, in Simizu B, Terasima T (eds): Vero Cells, Chiba University, Japan, 1987, pp 26-29.

4 ATCC Catalog – ATCC CCL 81 Vero, 1979, 54.

5 Montagnon BJ, Fanget B, Nicolas AJ: The large scale cultivation of Vero cells in micro-carrier culture for virus vaccine production. Preliminary results for killed poliovirus vaccine, in Hennessen W, Van Wezel AL (eds): Reassessment of Inactivated Poliomyelitis Vaccine. Dev Biol Stand. Basel, Karger 1981, vol 47, pp 55-64.

6 Montagnon B, Vincent Falquet JC, Fanget B: Thousand litre scale microcarrier culture of Vero cells for killed polio virus vaccine gives promising results, in Griffiths B, Horaud F, Spier R, Hennessen W (eds): Fifth General Meeting of European Society of Animal Cell Technology. Dev Biol Stand. Basel, Karger 1984, vol 55, pp 37-42.

7 Montagnon BJ, Fanget B, Vincent Falquet JC: Industrial-scale production of inactivated poliovirus vaccine prepared by culture of Vero cells on microcarrier. Reviews of Infectious Diseases, 1984, vol 6, suppl.2, pp S341-S344.

8 Montagnon BJ: Polio and rabies vaccines produced in continuous cell lines: a reality for Vero cell line, in Hayflick L, Hennessen W (eds): Continuous Cell Lines as Substrates for Biologicals. Dev Biol Stand. Basel, Karger, 1989, vol 70, pp 27-47.

9 Van Steenis G, van Wezel AL: Use of the ATG-treated newborn rat for in vivo tumorigenicity testing of cell substrates, in Griffiths JB, Hennessen W, Horaud F, Spier RE (eds): The Use of Heteroploid and other Cell Substrates for the Production of Biologicals. Dev Biol Stand. Basel, Karger, 1982, vol 50, pp 37-46.

10 Carcagne J, Ha KT, Armand J: Evaluation of transforming activity of cellular DNAs from different origins by NIH 3T3 transfection test. Biologicals 1991;19:317-325.

11 Waldeck W, Sauer G: New oncogenic papovavirus from primate cells. Nature (London) 1977; 269:171-173.

12 Montagnon BJ, Martin R, Barraud B, Bandet R, Fanget B, Paturel J, Mackowiak JF: Residual Vero cell DNA and inactivated polio vaccine : detection by DNA-DNA molecular hybridization, in Hopps HE, Petricciani JC (eds): In Vitro-Cellular and Developmental Biology: Abnormal cCells, new Products and Risk, 1985, Monograph Number 6, pp 82-89.

13 Horaud F: Absence of viral sequences in the WHO-Vero cell bank. A collaborative study, in Brown F, Esber EE, Williams M (eds): Continuous Cell Lines – An International Workshop on Current Issues. Dev Biol Stand, Basel, Karger 1992, vol 76, pp 43-46.

Dr. B.J. Montagnon, Pasteur Mérieux Sérums & Vaccins, 1541, av. Marcel-Mérieux,
F-69280 Marcy l'Etoile, France.

Brown F, Griffiths E, Horaud F, Petricciani JC (eds): Safety of Biological Products Prepared from Mammalian Cell Culture.
Dev Biol Stand. Basel, Karger, 1998, vol 93, pp 125-128

DNA in Plasma of Human Blood for Transfusion

M. Duxbury, M. Jezuit, B. Letwin, J. Wright

Genetics Institute, Cambridge, MA, USA

INTRODUCTION

When continuous cell lines were first used for the production of biologicals in the early 1980s [1], concern over potential contaminating cellular DNA initially resulted in acceptable levels being proposed at 10pg/dose [2]. This was later modified to 100pg/dose based on the suggestions of a WHO Study Group [3]. Most of this attention focussed on manufactured products such as monoclonal antibodies, vaccines, and proteins derived from recombinant DNA technology. However, if contaminating DNA is a valid safety concern, then normal human DNA, which contains protoncogenes, in products such as whole blood used for transfusion, also carry a theoretical risk. This study was designed to estimate the levels of human DNA that may be released on the death of nucleated blood cells during the normal 21 day storage of human blood, and thus to compare them to the proposed acceptable limits of other biologicals.

MATERIALS AND METHODS

Preparation of Plasma Samples

Seven expired units of non-irradiated, HIV- and hepatitis-antibody free whole blood and stored in CPD (citrate, phosphate, dextrose), were purchased from a local hospital bank. The units (two type O Rh+ve, one type O Rh-ve, and five type A Rh+ve) were each mixed by inversion and aliquots were centrifuged at 2000 rpm for 25 min at 4°C. The resulting plasma was stored for up to six weeks at 4°C until tested.

To examine the possible changes in contaminating DNA levels during the storage, whole blood was drawn from three healthy individuals, and stored in CPD at 4-8°C. After one, four, seven, and 21 days, plasma was prepared from whole blood aliquots as described previously and stored at -80°C until tested.

Estimation of DNA Content

DNA content was estimated using the Threshold™ Total DNA Assay Kit (Molecular Devices Corporation, Menlo Park, CA, USA). Samples were diluted as appropriate in sterile zero calibrator solution from the kit followed by overnight incubation at 60°C in the presence of 100 µg/mL proteinase K or 500 µg/mL pronase (Boehringer Mannheim, Indianapolis, IN, USA) plus 100 µg/mL proteinase K to digest the plasma proteins. Duplicate 0.5 ml samples were also prepared, each containing 50 pg of calf thymus DNA to monitor accuracy of quantification. Samples and standards (200 to

3.1 pg/0.5 ml) were heat-denatured (100°C, 15 mins.) and cooled rapidly to 0°C. The DNA content of the prepared samples was then estimated according to the recommended test protocol.

Results

The results of the analysis of seven units of expired whole blood are shown in Table 1. All recoveries of the 50pg DNA spike into test sample were acceptable. Three samples were analysed at two different initial dilutions, but after adjustment for dilution the results showed reasonable agreement. These data confirm that the plasma matrix did not interfere with the estimation of DNA and thus the reported estimates can be considered accurate. The results suggest that after 21 days storage, human whole blood can contain 250-1500 ng/ml released DNA. This means that the total amount of DNA that may be administered to a patient during the transfusion of one unit of whole blood may be as much as 450 µg (based in 1.5 µg/ml DNA and 60% plasma volume of a 500 ml unit of whole blood).

The results of the analysis of DNA in plasma from whole blood during 21 days storage indicate that there is a noticeable increase in DNA levels during the typical storage period (Fig. 1). The final levels in this experiment were much higher than detected in expired blood from the blood bank. However, the collection and storage of these samples did not exactly reproduce the optimal collection conditions used during normal blood donation and elevated levels of cell death and thus released DNA would not be unexpected.

Discussion

The Threshold™ Total DNA Assay has been shown to be capable of measuring the level of DNA from various sources [4]. It uses a two-site binding approach

Table 1: Results of DNA analysis of seven samples of plasma from expired human donor blood. Mean level estimated = 634 ng/ml (Range 260-1474 ng/ml).

Unit number	Dilution	Corrected ng DNA/ml plasma *average		% Recovery 50pg. DNA spike
34999	5000	284	260*	98
	10000	236		102
35010	5000	324	283*	121
	10000	242		112
35018	5000	503	526*	77
	10000	548		92
35022	5000	624		68
35079	5000	354		107
35088	5000	1474		99
35100	5000	915		95

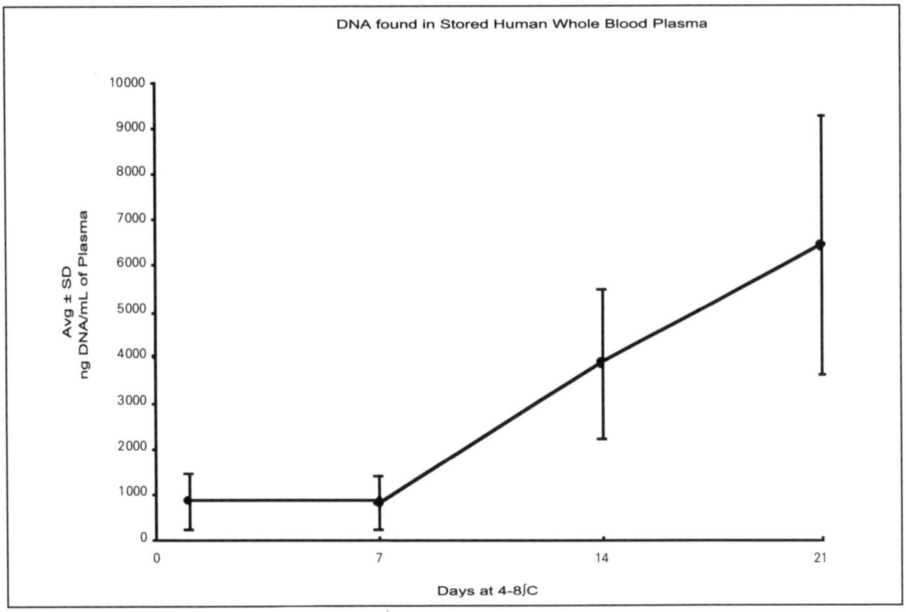

Fig. 1: Increase in DNA levels during storage.

which results in DNA of less than 800 bases giving a reduced response [5]. Despite this limitation, it would appear that as much as 450 µg of DNA greater than 800 base pairs could be administered during a routine transfusion of a unit of human whole blood.

While this study is limited to seven units of expired blood and three samples analysed during a storage time course, the results suggest that contaminating cellular DNA is in fact at significantly higher levels in transfused blood than the limits proposed for other biologicals, including those manufactured using recombinant DNA techniques.

Studies have shown a 25% loss of average nuclear count in stored human blood after 16-20 days [6]. Assuming 10pg DNA/cell and 8 x 10^6 white blood cells/ml of blood, then 25% degradation would result in the release of approximately 2 x 10^4ng cellular DNA/ml. These levels are consistent with the data presented here. Similar studies have shown that chromosomal alterations can occur during storage of whole blood [6]. These data suggest that DNA present in the plasma of stored blood could be rearranged, thus increasing the likelihood of the presence of oncogenes. However, there is no evidence to suggest that such levels of contaminating DNA in whole blood carries any risk. Indeed, there appears to be a much greater risk from incompatibility and potential contamination with infectious agents. Furthermore, the level of contaminating cellular DNA in transfused whole blood is five to six orders of magnitude greater than the 100 pg/dose limit recommended by the WHO. In view of the lack of evidence of any risk associated with these levels of cellular DNA contamination in whole blood, further consideration should be given to revising the need to reduce DNA in other biologicals to such low levels.

References

1 Petricciani JC, Salk PL, Salk J, Noguchi PD: Theoretical considerations and practical concerns regarding the use of continuous cell lines in the production of biologicals, in Griffiths JB, Hennessen W, Horaud F, Spier RE (eds): The Use of Heteroploid and other Cell Substrates for the Production of Biologicals. Dev Biol Stand. Basel, Karger, 1982, vol 50, pp 15-25.

2 Hopps HE, Petricciani JC (eds): Abnormal Cells, new Products, and Risk. Gaithersburg, MD, USA, Tissue Culture Association, 1985.

3 Petricciani JC, Hennessen W (eds): Cells, Products, Safety. Dev Biol Stand. Basel, Karger, 1987.

4 Kung KT, Panfili PR, Sheldon EL, King RS, Nagaini PA, Gomez B Jr, Ross DA, Briggs J, Zuk RF: Picogram quantitation of total DNA using DNA binding proteins in a silicon sensor based system. Anal Became 1990;187:220-227.

5 Briggs J, Painful PR: Quantitation of DNA and protein impurities in biopharmaceuticals. Anal Chem 1991;63:850-859.

6 Turner JH, Hutchison DL, Petricciani JC: Cytogenetic and growth characteristics of human lymphocytes derived from stored donor blood packs. Scand J Haemat 1971;8:169-176.

Dr. M. Duxbury, Genetics Institute, 87, Cambridge Park Drive, Cambridge, MA 02140, USA.

Brown F, Griffiths E, Horaud F, Petricciani JC (eds): Safety of Biological Products Prepared from Mammalian Cell Culture.
Dev Biol Stand. Basel, Karger, 1998, vol 93, pp 129-140

Discussions

Detecting Viruses in Cell Banks

Chairperson: P. Minor
Rapporteur: W. Berthold

The presence of SV40 DNA sequences has been reported in several types of tumours in children and adults, including mesothelioma. It has been speculated that the source is polio vaccine, and it is known that vaccines containing SV40 were used in the late 1950s. The appearance of SV40 is not a cell-bank incident: the primary monkey kidney cells used to produce the vaccine were obtained from rhesus monkeys, which may have been infected. The source was then changed, replacing rhesus by African cereopithecus and vervet monkeys, which are not contaminated. SV40 may be present in inactivated poliovirus vaccine because it is much more resistant to inactivation than poliovirus. An epidemiological study of a group of children receiving vaccine containing SV40 in the USA, however, has shown no difference in the rates of cancer. Oral and inactivated poliovirus vaccines produced from the early 1960s are free of SV40. Even if it is found that SV40 is not pathogenic for man, SV40 cannot be accepted as a contaminant of cell banks. SV40 sequences have also been detected in tumours of young children, indicating sources other than the contaminated poliovirus vaccines of the 1950s and 1960s.

There are two aspects to the problem: public health, which is the real problem, and public confidence. Every time SV40 contamination is found, vaccination programmes receive a setback. A useful epidemiological approach might be to examine biopsy samples from patients of different ages with mesothelioma in different geographical areas; alternatively, samples of tumour material embedded in paraffin wax could be used for SV40 amplification. Comparisons could then be made with good tumour registries in an international study.

Many continuous cell lines contain retroviruses, and the altered genes have been there for many years. Fibroblasts have the same genes but do not express them. It was concluded that cell banks should be specifically virus-free, but the list of specific viruses would be a long one and it would have to be decided which to incorporate in the list.

In-process Testing for Viral Agents

Chairperson: P. Minor
Rapporteur: W. Berthold

Several kinds of test have been developed, including in vitro cell culture and in-vivo methods in animals and eggs.

J. Robertson described the reverse transcriptase (RTase) assay, in which primers are introduced with target RNA: when target RNA is converted into cDNA, the test sample is assumed to contain RTase. New methods are much more sensitive than those used previously. Vaccines derived from chick cells contain some transcriptase activities, implying that unknown retroviruses could be present in certain live attenuated vaccines, such as those against measles, mumps, yellow fever and influenza. The transmissibility of the activity was investigated in a range of cell types including human diploid and human lymphoid-derived cell lines, and in avian cells, which are permissive for avian retroviruses. No evidence was found for the presence of retroviruses. The activity is filterable and can be found in a sucrose-density gradient at a density similar to that of typical avian retroviruses; however, electron microscopy reveals no characteristic retroviral particles. Attempts at elucidating the nucleic acid sequence have not met with success. RTase assays are commonly used in testing for retroviruses, but they are so sensitive that they can detect levels of enzyme activity that are of no relevance to infection, indicating that the assay would not be useful in ensuring safety. The difference in RTase activity between an infected and an uninfected avian cell might be so great, however, that a cut-off limit could be assigned to the level of background activity of a conserved element as opposed to the presence of an actual infectious virus.

Mammalian cells, such as MRC-5, that have been used as indicator cell lines were negative in the assay. The conditions of the assay might result in a negative result with human diploid cell lines, whereas if the parameters, such as types of ions and the pH of the RTase reaction, were altered, a positive result might be obtained.

Of all the assays that would ensure safety, most of those used to establish the acceptability of cell lines are infectivity assays. Other assays, such as immunological and PCR assays, are sensitive but can be used only to indicate further investigation of any biological activity detected. In comparative studies, the results depend on the assay used. A coded sample may be positive in one assay and negative in another. The conditions are being examined carefully to see how the sensitivity could be improved. If progress is to be made, a standardized assay should be chosen that gives consistent results. Detecting the possible presence of retroviruses is an important aspect of safety evaluation, and the most sensitive test for retroviruses should be used, even if positive results do not always indicate the presence of an infectious virus. Thus, a positive result should be followed by an investigation of the basis of that positivity. International collaborative studies should be conducted to establish consistency between laboratories, notwithstanding the different methods in use. Some reference materials might be developed so that agreement can be reached on a cut-off value for infectivity.

In the USA, hepatitis B virus (HBV) vaccine is produced in genetically modified yeast and mammalian cells which express the surface antigen, HBsAg. Hepatitis B kills 1.1 million people in the world every year. Chronic HBV infection is detected by the presence of HBsAg in the blood and some integrated HBV DNA might be found in the liver. The small surface antigen is currently used as a vaccine. Mutants may occur under immune selection pressure: in a highly endemic environment such as Taiwan, southern Europe and Italy, the clinical relevance of mutants is not known. Mutant virus antigens could be included in future vaccines, if needed.

One solution to the presence of contaminating viruses in a cell culture used for the production of biologicals is a well-characterized cell bank. The current state of the art of characterization is the use of in vitro or in vivo methods, or a combination of both. Adventitious contamination by endogenous viruses that replicate in the later stages of the production cycle has been addressed by a number of manufacturers. The question that remains is the level and timing of control. The major concerns include specific inactivation steps for viruses that were not expected to be present. The best time to test for viruses is at the end of production. General screening assays can be used to detect a broad range of viruses in a relatively short time and then specific assays for viruses that are considered to be a problem. Good results have been obtained with PCR, with a false-positive rate for any specific virus of virtually zero. This test takes about 8 h, whereas a cell culture process can take 7-14 days but may reveal a number of viruses. Minute virus of mice (MVM) is an adventitious virus that can enter the system from an unknown source and has been known to proliferate in CHO cells. It is an excellent model virus for validating the recovery process. A relatively inexpensive, easy multiple PCR assay is being developed to detect many rodent viruses. As the assay detects DNA and not the presence of an infectious virus, however, an infectivity assay will be included. This test was to be developed primarily because of its speed. In a test that takes about one month, contamination might have inadvertently spread throughout a facility. If contamination can be detected in a single culture before breaking containment, only that culture will be lost.

Testing Source Materials

Chairperson: J. Peetermans
Rapporteur: R. Wolff

R. Wolff emphasized two aspects for discussion: the animals that serve as the source of cell substrates and animal and human-derived culture components, including sera and trypsin. Ideally, sources should be available that are free of risk for transmission of pathogenic organisms. In the real world, that is not possible. The points for discussion were the kinds of tests to be conducted on animals used to generate cells for production substrates, on herds used to generate cell culture components and on cell culture components, the size of the sample for non-homogeneous distributions, the design of the test system and the susceptibility of production substrates; and, within production, how manufacturers could set up barriers for transmission of agents.

If cell cultures from monkeys are used, monkey viral contamination is the main problem. The current recommendation is that monkeys should be tested for antibodies to SV40, simian retroviruses or SIV. SV40 has never been found in some wild monkey strains and seems to occur only in rhesus monkeys; these should therefore be excluded from colonies and animal facilities. Chicken flocks are tested for about 25 agents by serological monitoring. The best solution to the problem of primary sources would be to replace them by adequate cell lines that can be banked.

Live viral vaccines are difficult to purify. The same testing as done on continuous cell lines is also done on live viral vaccine control cells and also on the vaccine itself, during each production run. If one is using the only cells in which a product can be made, the presence of virus might be acceptable if it were proven to be cleared during processing. If cells that are free of virus are available, however, there is no reason to use cells in which virus is known or suspected to occur. Each barrier to contamination has its limitations. The results are related to the sampling procedures and the sensitivity of the tests. If PCR were to become a routine part of testing as opposed to depending on cell culture assays, it might be too sensitive and pick up irrelevant signals. Several live vaccines are extremely effective and are reducing major infectious diseases throughout the world. Thus there are good reasons for not radically redesigning them.

The presence of viruses should be investigated in porcine and bovine trypsin. Bovine serum is tested for the absence of bovine viruses by testing comparable to that done for cell banks and for primary control cells. The other solution would be to replace bovine supplements in cell cultures or to grow continuous cell lines in the absence of serum. A large amount of serum is needed to grow MRC-5 cells and all primary cells. In principle, all material of animal origin should be inactivated or autoclaved, instead of using fresh material that has simply been filtered. However, it would be impossible in the short or medium time to sterilise or replace all products of animal origin with synthetic or vegetable products. However, such radical virocidal treatment may at the same time also remove all the desired qualities of this raw material for which it was chosen in the first place.

Many products contain a well-known product of human origin as an excipient or stabilizer, namely human albumin. The production process and the final pasteurisation will eliminate a wide variety of infectious agents. However, some regulatory agencies withdraw human albumin if it is shown subsequently that one of the donors from which it was derived, or even a member of the family, developed Creutzfeld-Jacob disease. There may be no real risk, but the virtual risk, the public perception of safety, is important. The sooner use of such human and animal material in vaccines can be avoided, the better. Unfortunately, human serum albumin is such a highly active substance for stabilization that is would be very difficult to replace. A study of replacements should be begun immediately. A recombinant product is already available. Albumin is made from very large pools of donors, with a high likelihood that one of the donors will later die of an infectious disease. Pools of a smaller number of donors, under the age of 50 years of age, could be used and tested carefully by PCR, for example, for likely contaminating agents.

Monkey kidney cells are still used for the production of OPV, so that great care should be taken before changing any of the established recommendations, which would have implications for the WHO expanded programme of vaccination. The meeting could recommend to WHO that an expert group specifically review the production of polio vaccine in primary monkey cells, taking into consideration new techniques.

Hundreds of animal-derived products are used in fermentation, including simple products like lactose. It would therefore not be easy to avoid animal-derived materials.

The general consensus was that WHO should consult with experts to redirect some vaccine production that is currently based on primary cells. The processes of cell culture should be examined to redirect animal-derived materials to better-defined materials carrying less risk.

Bovine Spongiform Encephalopathy and Other Non-Viral Transmissible Agents

Chairperson: P. Brown
Rapporteur: L. Wudl

P. Brown said that little is known about these agents, other than that they contain no disease-specific nucleic acid; however, the prion protein is inseparable from infectivity. It has a molecular mass of about 30 kDa, and its three-dimensional structure resembles that which had been predicted. It appears to be a helically twisted double fibril, called a prion rod or a scrapie-associated fibril, similar to the structures seen in the brains of patients with Alzheimer's disease. However, as a soluble fraction can be isolated from infectious brains that does not contain fibrils, they are almost certainly an artefact of extraction. They probably resulted from the use of detergents in extracting and purifying tissue. In the available tests, 10 000-100 000 infectious units are needed to detect the protein. Biological assays remain the gold standard for screening biological products, but it would be impossible to check every biological product for the protein.

A test is available for assaying the cerebrospinal fluid (CSF) of affected people, which is useful for diagnosing the disease in humans, but it is not yet known if it can be used for diagnosing the analogous diseases in animals. Unfortunately, it is not specific, and the agent that is measured is simply a product released by dying neurons. Fortunately for diagnosis, the neuronal destruction in most cases of CJD exceeds the threshold and can be detected, but the test is of no use in the context of biological products. The epidemic in cows will die out, but scrapie has been around for at least 300 years and will probably exist forever. An experiment is being undertaken to test serial spinal fluid specimens from preclinically infected sheep.

R. Bradley reported a test by two-dimensional electrophoresis, which is specific for two proteins released into the CSF as a result of nonspecific neuronal damage. It is not effective in detecting infected animals before neurodegeneration occurs but allows a diagnosis of positivity, which might make it possible to save the remaining herds in the United Kingdom. Legally, brain tissue samples for research purposes could not be collected from live animals but only *post mortem*; however, if the CSF test were validated, samples could be taken in the field for diagnosis. In a test in hamsters, neuronal damage can be detected 30 days after intracerebral inoculation. Early signs of disease are seen after about 54 days, and death occurs after 60-65 days.

Manufacturers are using bovine materials from countries where there is no evidence of BSE, but soon there may be no 'BSE-free' countries. What are the processes by which potentially infected materials might be inactivated? Tests are available that clearly distinguish between native prion and altered infected prion, as a treatment exists that destroys the normal, non-infectious form.

In the case of human disease and human blood and blood products, epidemiological surveillance programmes are being conducted. Of the cases of CJD in the European Union, about one-quarter had family members who were blood donors, and about one in 10 CJD patients had donated blood. In the USA, there had been a huge recall of plasma because someone who had given blood over a 10-year period up through the initial clinical stages of CJD had died. One or two cases a year will not, however, provide the statistical power necessary for significance in the study of CJD patients who have and have not received blood. In retrospective studies, attempts are being made to identify all of the recipients of blood from a person who later died of CJD, but 100% of the donors had not been identified. It will be important to resolve the problem epidemiologically within the next few years.

A firm producing material that is a mixture of relatively low molecular mass polypeptides from pig brain for use in the therapy of Alzheimer's disease and other neurological diseases uses several steps to remove or inactivate agents, including exposure to an organic solvent and a terminal filtration through a 10-kDa molecular sieve. Latent CJD or BSE agent could not be recognized in the pigs because they are slaughtered at the age of one or two years, before the disease appears. The agents can be inactivated fairly reliably by processes that are absolutely incompatible with the biological activity of most products, e.g. autoclaving for 1 h at 134°C or exposure to 1 N sodium hydroxide or sodium hypochlorite. Two gentle but unpredictable means of removing a large amount of infectivity are chromatography and filtration; the former could be expensive and the latter means that the product must be small. Using these procedures, rubella and AIDS viruses were inactivated, but PrP was not, even after exposure to 1 N sodium hydroxide.

The conclusions were that there are no assays that are sensitive or specific enough to test raw materials or sources. There is a great deal of concern about excipients and especially human albumin. A number of barriers could be used to exclude prions, even in the finished biological product. One is to ensure a clean source, so that bovine products used as raw materials are derived from tissues that have been demonstrated to be non-infectious. When possible, raw materials should be autoclaved or inactivated by some method, and some inactivating steps should be included in the downstream process, with or without a 10-kDa barrier filter, including chromatography resins to remove prion protein nonspecifically by adsorption or specifically by protein fractionation.

Karyology

Chairperson: R. Hay
Rapporteur: W.W. Nichols

J. Petricciani emphasized the importance of differentiating the use of karyology for chacterizing a new diploid cell line and its routine use as a quality control measure. The rationales used for setting guidelines in the 1960s and 1970s should be reviewed to see whether they continue to be valid in the 1990s and why some are applicable only to diploid cells and not to primary cells and continuous cell lines. W. Nichols recalled that all of the current fears about using human diploid cells had also been present previously. The objective of using karyology is to detect spontaneous or induced transformation, to detect contamination, to determine cell identity, and to screen nonspecifically for viral infections, since 20-30 viruses have been described that increase chromosomal breaks and abnormalities. A very conservative approach was used initially, but by 1978 it was recognized that the guidelines should be made less restrictive. Neither MRC-5 nor WI38 cells nor any normal diploid strain transforms spontaneously. Cells from patients with Li-Fraumeni syndrome, a hereditary cancer-susceptibility syndrome in which there is a mutant *p53* gene, become immortalized but they cannot produce tumours. Cells that transform show remarkable changes in morphology and growth characteristics, such that cytogenetics are not needed. Contamination with a heteroploid line can also be identified by growth characteristics and morphology. Molecular techniques are much more efficient for distinguishing one diploid cell strain of the same sex from another. Short tandem repeat analysis could be used. Two multiplex commercial kits are available for identifying a cell line, with an error rate of 1 in 350 000. DNA fingerprinting detects only about 10% of a contaminating line. Two cell lines do not always undergo mitosis at the same frequency; therefore, a 1% contamination might be represented by only 0.01% mitotic figures.

If a human diploid cell is considered to be intrinsically safer than a continuous cell line, all that is necessary is to prove that the former is a human diploid cell and is not contaminated with a continuous cell line. Contamination of a continuous cell line with another continuous cell line gives cause for concern in research and from a manufacturer's point of view, as it would indicate that two heteroploid lines were growing together. As heteroploid lines change chromosomally throughout their lifespan, they cannot be characterized like diploid lines. Other cell lines, such as BHK 21, CHO and Vero, which might be used in a manufacturing process, should be characterized early in their use. No two CHO karyotypes are alike, but they have the same cell line-specific markers. Continuous cell lines can be a risk because they allow growth of some human viruses. The possibility of heavy contamination of a diploid cell process is low.

Characterization does not rule out contamination; it shows how cells perform with time and whether there is spontaneous breakage of chromosomes earlier than expected. Contamination with a tumour cell can be seen readily under the microscope.

Is karyology useful in routine quality control of production runs with heterodiploid cell lines? The consensus appeared to be that it is not required for routine monitoring of the population. It could be used as an identity test at the end of production. MRC-5 and WI38 cells are a special case. They have been used for growing vaccines successfully for about 30 years. Karyology has been used for detection of contamination with heteroploids, spontaneous transformation and adventitious agents. The current numerical specifications were established in 1979 on the basis of data for WI38 cells in conventionally stained preparations and not new techniques such as banding. Diploid cells are far superior to primary tissue, but the numerical specifications of passage number might mean that irreplacable material would have to be discarded. Routine chromosomal characterization and monitoring and tumorigenicity testing are therefore not required for WI38 or MRC-5. The WI38 and MRC-5 cultures had been used for many years, and millions of doses of vaccine had characteristic features. What had changed was that more and better information was available about the range of abnormalities and their frequencies. At any one time, each person has about six million lymphocytes with translocations. Every diploid cell line that has ever been established, from normal adults, foetuses, skin or lung, all have translocations and clonal events, which are not related to loss of senescence or to tumorigenicity. The limits of acceptability that were established for WI38 cells are irrelevant even within the same foetus, as there is much heterogeneity even within the same individual.

Cytogenetics are therefore not needed routinely for WI38 and MRC 5 cells. A new diploid cell line should undergo chromosomal characterization, but the requirements could be made less restrictive: instead of requiring 1000 karyotypes four times during the life span of a cell, several hundred karyotypes four times in several laboratories would suffice. On every production run using heterodiploid cells, the identity of the cell type should be tested, to distinguish between a human diploid cell and a continuous cell line. Karyology or a DNA forensic test with a PCR or Southern blot method could be used.

The anomalies considered acceptable previously have changed with the advent of banding techniques. Most structural rearrangements could not be recognized before banding; nor could translocations unless they changed the size and centromeric position. With banding, a clonal event can be identified. A translocation or a clonal event is not lethal, but most chromosomal abnormalities kill the cell. Rarely, a structural abnormality produces a tumour cell, which takes over the culture. The acceptable level of hyperdiploid cells is currently too low. Lifespan studies are needed for the karyology of a new strain.

A manufacturer using a new diploid cell line could assume that 200 metaphases had been studied in two laboratories at four periods through the cycle. He would then have to look for virological and endogenous agents, such as micoplasmas. A tumorigenicity study should also be done.

The NISH substrate group emphasized that cellular DNA levels must be consistent with acceptable levels. The arguments for treating diploid cells differently from others include the quantity of DNA in the product: those made of human diploid cell lines ought to be permitted to contain more DNA. The cell bank in the United States had well in excess of the proposed 500-ng limit. It was agreed that the guidelines should be made and should not be too rigid.

Karyology should be adapted to newer techniques. It should still be used in characterizing new diploid cell lines, but is not necessary for WI38 or MRC 5 cells. Lifespan karyology and immunogenicity studies should still be done, and some limited studies should be done on heteroploid lines. The guidelines for use of karyology in characterization should be modified to require fewer cells and a minimum number of changes.

DNA

Chairperson: G. Schild
Rapporteur: I.G. Furminger

G. Schild suggested that the discussion should cover a wide range of substrates, including prokaryotic cells, primary and secondary cell cultures of either human, simian or avian origin and other potential sources of human diploid cells and transformed cell lines. It should also cover a wide range of different products, including vaccines used in millions of healthy individuals and live vehicles for gene therapy. A new element was vaccines specifically composed of DNA. The group should also discuss whether the 1987 WHO expert group recommendation of an upper limit of 10 pg of DNA per dose of any product generated from transformed cells should be modified.

Tumorigenicity testing could be done in animals and in vitro. A classical distinction should be made between products «purified» with regard to DNA, cells, replication-competent cells and protein. No tumorigenicity testing is needed for cells used in the preparation of purified products or on MRC 5 or WI 38 cells; however, new diploid cells require testing. An «unknown» cell line might be one that was previously not licensed or used. There were two positions about the use of crude, unpurified, possibly cell-containing material: never to use them as crude products or to decide on a case-by-case basis. Tumorigenicity assays have been looked upon as essentially negative assays, but they might be useful for characterizing the tumour type induced and for determining risk factors associated with the DNA or other components of a cell type. The way a dose-response curve evolves indicates the tumour-forming capacity of a cell and the risk factors associated with the DNA in that cell line. If the tumour-inducing capacity of a cell were known and related to the quantity of DNA, the level of risk associated with various concentrations of DNA in any product could be estimated.

There appeared to be a broad measure of support for knowing about the tumorigenic properties of cells used to make products, particularly those to be used in large numbers of healthy individuals. If the product were highly purified, tumorigenicity would not necessarily exclude use of the cell line. Application of tumorigenicity studies would thus be on a case-by-case basis, depending on the nature of the cell and of the product and its degree of purity.

R. Weiss remarked that cancer researchers studied the specific sequences in DNA that were altered in different tumour cells. It might be important to establish the genetic basis for the tumorigenicity and the kinetics and dosage.

Assays of cell-derived nucleic acid in biological products could contribute to knowledge about safety or consistency or might simply indicate purity. The original decision about a safe level of DNA was based mainly on studies in animals with DNA isolated from viruses. Should the requirements be applied to products made in prokaryotic cells, yeast cells, human diploid cells or transformed cells? Is there a scientific basis for changing the recommended level of DNA?

The original document concluded that a specific risk was associated with heterogeneous cells derived from continuous cell lines, but continuous cell line products now contain no DNA. DNA obviously is present in DNA vaccines and gene therapy products. What must be avoided was insertion of DNA into chromosomes and insertional mutagenesis. In an insertional mutation, a protooncogene might be switched on or a suppressor gene switched off; alternatively, an oncogene present in the DNA contaminating the product might be inserted and expressed. If a good assay were available for investigating the tumorigenic potential of contaminating DNA, there would be less concern about indirect assays of karyology, in-vivo tumorigenicity and quantities of DNA, the meaning and value of which are not known. In a collaborative study, the ability to detect DNA had been highly variable, and false-positive results had been obtained, depending primarily on the process used to extract pure DNA.

The value of 100 pg is now 10 years old. It is based on data for SV40 and polyoma and not cellular DNA. In a study in monkeys given DNA from human tumour T24 cells intramuscularly, intracerebrally or intravenously, there was no gross evidence of effects 10 years later. The question is whether the result can be extrapolated to the potential injection of a vaccine into millions of humans.

In a study of blood from a local blood bank, the amount of DNA increased over time, probably because of lysis of white cells, to 50-450 μg per unit. Plasma samples from oncology patients contain 25-50 μg of DNA per unit, and the possibility of using that DNA for diagnosing various types of cancer is being explored, as some would carry tumour markers. Donors might thus be in a premaligant state, and blood for tranfusion is therefore a source of potentially oncogenic DNA. DNA in blood is generally derived from cells that have undergone apotosis and is therefore only about 200 base pairs long; it therefore might not be expected to contain entire active oncogenes.

A major concern with regard to biological products and cell lines is still the infectivity of cellular DNA. One can calculate that the total risk of disease is 1 in 240 000. The caveats are (i) the DNA size distribution; (ii) the number of provirus copies per genome; (iii) human cell susceptibility to a specific provirus; and (iv) the difference between injected and oral products. The highest level of concern is associated with DNA from human retrovirus-infected cells. Second would come hybridomas, which also produce retroviruses and can infect human cells; they integrate provirus, and a hybridoma cell has multiple opportunities for reintegration of provirus. Chinese hamster ovary cells probably represent a smaller risk, but it is not known whether they contain provirus. A clearly non-tumorigenic continuous cell line like Vero cells is of less concern, and human diploid cell lines are of least concern. The ranking of human tumour cell lines depends on the line: some contain integrated viruses, including retroviruses, and individual lines would have to be characterized.

Small amounts of DNA might not be immunogenic, whereas larger amounts might be, depending on the ages of the people to whom a vaccine is given and on the adjuvants in the vaccine. Furthermore, although the quantity of viral DNA that can induce tumours is of the order of micrograms, the same sequence might not be tumorigenic under other conditions. The important aspect for a manufacturer is to be able to demonstrate that the purification technique is reproducible.

It was agreed that some number should be set, to guide regulatory authorities and to develop assays. In many processes, an extra step is added merely to reduce the DNA level to a set number. If the level of DNA were raised, manufacturers would save a great deal of money. Products should always, as a matter of principle, be as pure as possible. Manufacturers should be aware of the risks and be consistent in removal of DNA down to a certain level. The results of toxicological studies should be taken into consideration, allowing flexibility.

It might be premature to standardize assays and preparations for DNA testing, as it is not clear what form of DNA is of concern and various systems are used for making products; one should be cautious in interpreting the results of toxicological testing of products, as the results pertain to the toxicity of the molecule and not the contaminants.

Guidelines will presumably be drawn up for DNA vaccines, which contain various types of highly characterized DNA.

There was consensus that each batch of human diploid cells need not be tested for tumorigenicity but such tests are essential for new cell lines. Even if a cell line is tumorigenic, there are degrees of tumorigenicity, and products should be treated on a case-by-case basis. The purity of a product from a highly tumorigenic cell must be examined carefully. For live viral vaccines made on human diploid cells or chick embryo fibroblasts, there need be no limit on the DNA content, but other products should be considered on a case-by-case basis. The limit of 100 pg might be altered to 500 pg; alternatively, a range between 1 and 10 ng per dose might not be unreasonable.

Host-Cell Proteins

Chairperson: T. Hayakawa
Rapporteur: J. Robertson

J. Robertson noted that the products derived from primary and diploid cell lines are primarily vaccines with large quantities of impurities. Both these and product-related impurities, which might be variant viruses, might be immunogenic. The residual host-cell proteins from primary and diploid cell lines, such as cytokines, and some product-related impurities might also be reactogenic. Continuous cell line-derived products are typically highly purified and highly characterized and are used frequently; however, any impurities occur at such low levels that they are not immunogenic. Of greater concern are product-related contaminants, which might be immunogenic and inhibit the biological function of the molecule. The level of contaminating host-cell protein is small enough to rule out reactogenicity; the greatest concern is product-related immunogenicity and reactogenicity. Tests of host cell-derived proteins showed that they are not immunogenic at the levels at which specifications have been set. The major concern is de novo synthesis of antibody in the absence of pre-existing antibody.

All therapy with recombinant proteins carries a relative risk, as they all have impurities that cannot be eliminated. In most continuous cell lines, the impurities are below the level of biological significance. Recombinant products have been used in man for almost 10 years; probably more than one million people have received tissues and over 500 000 have received other products on a long-term basis. No significant clinical immunological problems have been encountered.

More attention should be paid to the spectrum of soluble factors produced by host substrate cells, including cytokines, lymphokines and chemokines. Some cell lines produce large quantities of tumour necrosis factor. Not only could they add to the undesirable features of a product, but various interleukins of rodent origin might be immunogenic and stimulate antibodies that react with the human counterpart molecule, which in turn may have undesirable consequences.

The impurities occur at much higher levels than DNA and are measured by very imprecise methods. Instead of measuring antibodies regularly throughout a process, which is extremely expensive, safety testing of the end-product should suffice. Protein and DNA impurities should be considered on a case-by-case basis and be relegated to process validation.

The value of having a number for any of the impurities instead of using a case-by-case approach or even a class basis is that it provides a target in the development of a process. If there is no target, the process is designed around fear of regulatory action, leading to excessive capital expense and waste of resources. The best way to improve human health is to allow more molecules to be tested rather than fewer.

It was concluded that product-related impurities rather than host-cell impurities are of concern. Product-related impurities should be considered in detail, including their potential to elicit an immune response against a naturally occurring protein. Industry had requested a hard number to aim for in setting process parameters and reliance on use of a validated process for protein impurities rather than developing and applying assays to every batch. There was general consensus that process validation could be used for DNA as well as cell proteins, rather than final product testing on each one.

G. Schild noted that there had been a general discussion about the actions that might be taken in the long term to avoid potential future safety issues. The desirability of avoiding use of primary or secondary cultures from animals had been raised and could be taken up by WHO. In the long term, it would be essential to avoid use of bovine or other animal materials, including human albumin. It had to be a long-term objective in order not to interrupt the supply of perfectly safe, good biological products. Perhaps recombinant products would be used as substitutes. Industry should become involved in such long-term research and development.

Brown F, Griffiths E, Horaud F, Petricciani JC (eds): Safety of Biological Products Prepared from Mammalian Cell Culture.
Dev Biol Stand. Basel, Karger, 1998, vol 93, pp 141-171

WHO Requirements for the Use of Animal Cells as in Vitro Substrates for the Production of Biologicals*

(Requirements for Biological Substances N° 50)

INTRODUCTION

Historically, the major concerns regarding the quality of biological products produced in animal cells have been related to the possible presence of adventitious contaminants and, in some cases, to the properties of the cells themselves. There are additional concerns regarding the quality of products prepared using recombinant DNA technology in relation to the expression construct contained in the cell substrates. It is well established that the properties of cell substrates and events linked to growth can affect the quality of the resultant biological products and, furthermore, that effective quality control of these products requires appropriate controls on all aspects of the handling of cell substrates.

GENERAL CONSIDERATIONS

Types of animal cell substrates

Primary cell substrates

Primary cells obtained directly from the trypsinized tissues of normal animals have played a prominent role in the development of virology as a science, and of immunology in particular. Cultures of primary cells from different sources have been in worldwide use for the production of live and inactivated viral vaccines for human use for more than 40 years, and experience has indicated that these products are safe and effective.

Major successes in the control of viral diseases, such as poliomyelitis, measles, mumps and rubella, were made possible through the wide use of vaccines prepared in primary cell cultures, including those from chicken embryos and the kidneys of

* Reproduced by permission of WHO from: WHO Expert Committee on Biological Standardization; forty-seventh report. Geneva, World Health Organization, 1998. (WHO Technical Report Series N° 878, pp 20-56).

monkeys, dogs, rabbits and hamsters, as well as other tissues. Cultures of monkey kidney cells have been used for the production of inactivated and oral poliomyelitis vaccines for more than 40 years, and the same cell system continues to be used for the production of both vaccines.

Primary cell cultures have the following advantages: they are comparatively easy to prepare using simple media and bovine sera; and they possess a broad sensitivity to different viruses, some of which are cytopathogenic. In addition, primary cells can now be grown in bioreactors using the microcarrier method [1].

However, where suitable alternative cell substrates are available, primary cell cultures are less likely to be used in the future for the following reasons: contamination by infectious agents, such as viruses, is a common problem; the quality and sensitivity of cultures obtained from different animals is variable; and it will become increasingly difficult to obtain cultures derived from nonhuman primates.

Primary cell cultures obtained from wild animals show a high frequency of viral contamination. For example, it is generally accepted that monkey-kidney cell cultures can be contaminated with one or more adventitious agents, including simian viruses. The number of viruses isolated and the frequency of isolation depend on many factors, including the method of isolation, test cell systems used, number of passages and duration of incubation and co-cultivation, and are directly proportional to the incubation period of the cultures. The frequency of contaminated cell cultures can be significantly reduced by careful screening of the source animals for the absence of antibodies to relevant viruses. The use of animals bred in a carefully controlled colony, especially those which are specific-pathogen free, is strongly recommended. The use of secondary or tertiary cells on which testing for adventitious agents can be performed will also reduce the frequency of contaminated production cell cultures.

Diploid cell substrates

The essential features of diploid cell lines of human (e.g. WI-38, MRC-5) or monkey (FRhL-2) origin are: they have a finite capacity for serial propagation, which ends in senescence; and they are non-tumorigenic and display diploid cytogenetic characteristics with a low frequency of chromosomal abnormalities of number and structure. Substantial experience over the past 25 years has been accumulated on the karyology of WI-38 and MRC-5 diploid cell lines, and ranges of expected frequencies of chromosomal abnormalities have been published [2]. More sophisticated cytogenetic techniques (e.g. banding) have demonstrated subtle chromosomal abnormalities that were previously undetectable, thus making the previously established ranges of abnormalities obsolete. Recent studies have shown that subpopulations of human diploid cells with such abnormalities may appear and disappear over time, and that they are non-tumorigenic and undergo senescence.

The possibility of using human diploid cell substrates for the production of viral vaccines was demonstrated more than 35 years ago. The experience gained with oral poliomyelitis and other viral vaccines in successfully immunizing millions of children in many countries has clearly demonstrated the safety of vaccines produced on such substrates [3].

The main advantage of diploid cell lines in comparison to primary cells is that they can be well characterized and standardized, and production can be based on a cell bank system. In addition, unlike the continuous cell lines discussed below, they possess a finite life and are not tumorigenic. The cell bank system usually consists of cell banks of defined passage levels and may include a master cell bank and a working cell bank.

However, diploid cell lines have the following disadvantages: they are not easy to use in large-scale production, such as bioreactor technology employing the microcarrier method; in general, they need a more demanding growth medium than other cell substrates; and they usually need larger quantities of bovine serum (either fetal or donor calf) for their growth than do continuous cell lines.

Continuous cell-line substrates

Continuous cell lines have the potential for an infinite life span and can usually be cultivated as attached cells or in suspension in a bioreactor. They have been derived by the following methods: (a) serial subcultivation of a primary cell culture of a human or animal tumour cell, such as HeLa or Namalva cells; (b) transformation of a normal cell having a finite life span with an oncogenic virus, for example, a B lymphocyte transformed by the Epstein-Barr virus; (c) serial subcultivation of a normal cell population generating a new cell population having an infinite life span; or (d) fusion between a myeloma cell and an antibody-producing B lymphocyte.

While cell transformation can occur spontaneously in various animal cells grown in vitro (continuous cell lines from African green monkey kidney cells (Vero), baby hamster kidney cells (BHK21) and Chinese hamster ovary cells (CHO) were established in this way), it has not been reported with human cells derived from normal tissues.

Hybridoma cells express monoclonal antibodies and hybridoma cell lines have generally been established from rodent hybridomas. Human hybridomas are obtained by the transformation of a B lymphocyte with Epstein-Barr virus, usually followed by subsequent fusion with a murine myeloma cell.

Continuous cell lines are now considered to be suitable substrates for the production of many biological medicinal substances and possess distinct advantages over primary and diploid cell substrates [4]. A cell bank system similar to that used for diploid cell lines provides a means for the production of biologicals for an indefinite period based on well characterized and standardized cells. Continuous cell lines tend to be less demanding than diploid cell lines; as a rule they grow well using ordinary media and serum, and some do not require serum at all. They can also be used in microcarrier cultures and/or suspension cultures for large-scale production in bioreactors.

However, many continuous cell lines express endogenous viruses and are tumorigenic. Their theoretical disadvantages therefore include the risk of tumorigenicity associated with residual cellular DNA that may encode transforming proteins.

In 1986, a WHO Study Group considered a number of issues associated with the acceptability of new cell substrates for the production of biologicals [5] and concluded that, in general, continuous cell lines were acceptable for this purpose, but that differences in the nature and characteristics of the products and in manufacturing processes must be taken into account when making a decision on the use of a particular continuous cell line in the manufacture of a given product. WHO Requirements for Continuous Cell Lines used for Biologicals Production were published in 1987 [6].

In addition, the WHO Study Group recommended the establishment of well characterized cell lines that would be of value to national control authorities and manufacturers of biologicals. In following up this recommendation, WHO developed a WHO master cell bank for Vero cells, a continuous cell line established from the kidneys of African green monkeys. The reason for selecting this cell line was that it offered the immediate prospect of improving the quantity and quality of several vaccines being produced in other systems.

A master cell bank of Vero cells was donated to WHO, by a manufacturer, at the 134th passage. The maximum passage level recommended for production is 150. Studies of tumorigenicity in newborn rats suggest that cells in the passage range 134-150 are not tumorigenic. Collaborative studies in 10 laboratories with respect to sterility, adventitious agents, tumorigenicity, presence of reverse transcriptase and identity showed that the WHO Vero cell bank met the WHO Requirements for Continuous Cell Lines used for Biologicals Production [6].

The WHO master cell bank of Vero cells is stored at the European Collection of Animal Cell Cultures (ECACC), Porton Down, England and the American Type Culture Collection (ATCC), Rockville, MD, USA. Producers of biologicals and national control authorities can obtain cultures of these Vero cells (free of charge), as well as additional background information, from Biologicals, World Health Organization, 1211 Geneva 27, Switzerland [7].

Potential risks associated with biologicals produced in animal cells

The main potential risks associated with the use of biologicals produced in animal cells are directly related to contaminants from the cells, and they fall into three categories: viruses and other transmissible agents; cellular DNA; and growth-promoting proteins. A summary of the risk assessment for each follows. More comprehensive statements have been published elsewhere on the risks associated with contaminating DNA [5, 8, 15] and growth-promoting proteins [5].

Viruses and other transmissible agents

The 1986 WHO Study Group reviewed the potential risk to human recipients of products manufactured in cells containing viral agents. These may include complete viruses with known patterns of replication, such as simian virus 40 (SV40), virus particles such as type A retroviruses, which can be visualized by electron microscopy, and persisting viral genomes or parts of genomes, for example those of the hepatitis B and Epstein-Barr viruses. As described below, cells differ with respect to their potential for carrying viral agents pathogenic for human beings.

Primary monkey-kidney cells have been used to produce hundreds of millions of doses of poliomyelitis vaccines over the past 40 years, and although latent viruses, such as simian virus 40, were discovered in these cells, control measures were introduced to eliminate the risk associated with the manufacture of vaccines in cells containing those endogenous viruses. Additional controls may be needed as new viral agents and technologies are identified.

Human and nonhuman primate lymphocytes and macrophages may carry latent viruses, such as herpesvirus and retroviruses. Continuous lines of non-haematogenous cells from human and nonhuman primates may contain viruses or have viral genes integrated into their DNA. In either case, virus expression may occur under in vitro culture conditions.

Avian tissues and cells harbour exogenous and endogenous retroviruses, but there is no evidence for transmission of disease to humans from products prepared using these substrates. For example, large quantities of yellow fever, measles and live influenza vaccines have been produced for many years in eggs that contain avian leukosis viruses, but there is no evidence that these products have had any harmful effects in their long history of use for human immunization.

Rodents harbour exogenous and endogenous retroviruses. Lymphocytic choriomeningitis virus and haemorrhagic fever viruses from rodents have caused disease in humans by direct infection.

Human diploid fibroblasts have been used for vaccine production for over 30 years, and although concern was initially expressed about the possibility of the cells containing a latent human virus, no evidence for such an agent has been found, and vaccines produced from this class of cell have proved to be free from viral contaminants.

In light of the differing potential of the various types of cells mentioned above for transmitting viruses pathogenic in humans, different types of testing are appropriate for products manufactured using these cells.

When either diploid cell lines or continuous cell lines are used for production, a cell bank system is used and the cell bank is characterized as specified in the appropriate requirements published by WHO. Additional methods such as testing for viral sequences or other viral markers should also be considered. Efforts to identify viruses constitute an important part of the characterization of cell banks.

When cell lines of rodent or avian origin are examined for the presence of viruses, the major emphasis in risk assessment is placed on the results of studies in which transmission to target cells or animals is attempted. Risk to human recipients should not be assessed solely on ultrastructural evidence of the presence of viral agents in the cells.

The overall manufacturing process, including the selection and testing of cells and source materials, any purification procedures used and tests on intermediate or final products, has to be such as to ensure the absence of detectable infectious virus in the final product.

There may be as yet undiscovered microbial agents for which there is no current evidence or means of detection. As such agents become identified, it will be important to re-examine cell systems for their presence. Positive findings will have to be discussed with the national control authority.

Cellular DNA

Primary and diploid cells have been used successfully and safely for many years for the production of viral vaccines, and the residual cellular DNA deriving from these cells has not been (and is not) considered to pose any risk. Continuous cell lines have an infinite life span due to the deregulation of genes that control growth. The DNA deriving from such cell lines is therefore considered to have the potential to confer the capacity for unregulated cell growth, or tumorigenic activity, upon other cells.

The 1986 WHO Study Group advised on the levels of contaminating DNA deriving from continuous cell lines used in the production of biologicals for human use [5]. Risk assessment based on an animal oncogene model suggested that in vivo exposure to one nanogram (ng) of cellular DNA, where 100 copies of an activated oncogene were present in the genome, would give rise to a transformational event once in 10^9 recipients [13]. On the basis of this and other available evidence, the Study Group concluded that the risk associated with residual continuous-cell-line DNA in a product is negligible when the amount of such DNA is 100 picograms (pg) or less per parenteral dose. In determining this limit, the perceived problem was not the DNA itself but rather minimizing the presence of specific DNA sequences coding for activated oncogenes.

Additional calculations suggest that the risk of insertional mutagenesis that could lead to a neoplastic event is extremely small. In one recent report, it was predicted that a 10-μg dose of DNA would result in the inactivation of two independent tumour-suppressor genes, by insertional mutagenesis, within a single cell of a vaccine recipient in only one of 10^7 recipients [9]. These very low calculated levels of risk are consistent with the limited human and animal experience to date [10, 16, 18].

Additional data published recently have shown that milligram amounts of DNA containing an activated oncogene from human tumour cells have not caused tumours in nonhuman primates during an evaluation period of 10 years [16]. Also, human blood contains substantial amount of DNA in plasma (75-450 μg per unit of blood) [19, 20]. Furthermore, contaminating DNA in a biological product generally occurs as small fragments unlikely to encode a functional gene.

The assessment of the safety of a product with respect to residual cellular DNA has to take into account: (a) the low levels of risk implied by the considerations described above; (b) the possible inactivation of any biological activity of contaminating DNA during processing; and (c) any reduction in the level of contaminating DNA during the purification process. A product may be considered safe on the basis of (b) and/or (c).

The current state of knowledge suggests that continuous-cell-line DNA can be considered as a cellular contaminant, rather than as a significant risk factor requiring removal to extremely low levels. On the basis of this reassessment, the Expert Committee concluded that levels of up to 10 ng per purified dose can now be considered acceptable. The purification process has to be validated by appropriate methods, including spiking studies, to demonstrate its capability to remove DNA to an acceptable level. In addition, batch-to-batch consistency needs to be shown for clinical trial batches and for three or more consecutive production batches. Subsequently, routine release testing for continuous-cell-line DNA in the final purified batch may not be needed. Any exceptions need to be agreed with the national control authority. For example, data suggest that β-propiolactone, a viral inactivating agent, may also destroy the biological activity of DNA; use of this agent therefore provides an additional level of confidence even when the amount of DNA per parenteral dose may be substantial [21]. Data should be obtained on the effects of such inactivating agents under specific manufacturing conditions so that firm conclusions on their DNA-inactivating potential for a given product can be drawn.

There may be instances where continuous-cell-line DNA is considered to pose a greater risk, e.g. where it could include infectious retroviral provirion sequences. Under these circumstances, acceptable limits should be set in consultation with the national control authority.

The new upper limit of 10 ng of residual DNA per dose does not apply to products derived from microbial, diploid or primary-cell-culture systems. The 1986 WHO Study Group stated that the risks for continuous-cell-line DNA should be considered negligible for preparations given orally; for such products, the principal requirement is the elimination of potentially contaminating viruses and toxic proteins. The upper limit of 10 ng of residual continuous-cell-line DNA per dose therefore does not apply to a product given orally. Acceptable limits should be set in consultation with the national control authority.

Growth-promoting proteins

Growth factors may be secreted by cells used to produce biologicals, but the risks from these substances are limited, since their growth-promoting effects are usually transient and reversible, they do not replicate, and many of them are rapidly inactivated in vivo. In exceptional circumstances, growth factors can contribute to oncogenesis, but even in these cases, the tumours apparently remain dependent upon continued administration of the growth factor. Therefore, the presence of known growth-factor contaminants at ordinary concentrations does not constitute a serious risk in the preparation of biological products from animal cells.

Proteins prepared using continuous-cell-line substrates need to be purified to permit their safe clinical use. Analytical methods to assure the purity of each batch should be proposed and validated by the manufacturer. The purification process should also be validated to demonstrate its capability to remove host-cell proteins to an acceptable level. In addition, batch-to-batch consistency should be shown for clinical trial batches and for three or more consecutive production batches. Subsequently, routine release testing for host cell proteins in the final purified batch may not be needed.

Requirements published by WHO

The first requirements published be WHO for cell cultures used for the production of biologicals were formulated in 1959 for the production of inactivated poliomyelitis vaccine in primary cell substrates [22]. They were revised in 1965 [23]. The successful use of primary cell cultures derived from the kidneys of clinically healthy monkeys for the production of both inactivated and oral poliomyelitis vaccine [24] led to confidence in the use of other cell cultures for the production of various viral vaccines. Many types of cell culture are now widely used for the production not only of viral vaccines, but also of other biologicals, such as monoclonal antibodies and a wide range of biologicals prepared using recombinant DNA technology.

Taking into account the latest available data relating to cell substrates and after extensive consultation, especially at a WHO/International Association of Biological Standardization/Mérieux Foundation International Symposium on the Safety of Biological Products prepared from Mammalian Cell Culture held in Annecy, France, in September 1996, the WHO Expert Committee on Biological Standardization adopted the text of this Annex as requirements appropriate for the quality control of animal cells used as in vitro substrates for the production of biologicals. They supersede previous requirements describing procedures for the growth and quality control of cell substrates for the production of biologicals [5, 6] and should be read in conjunction with the requirements published be WHO for individual products.

The following requirements concern the characterization and testing of continuous-cell-line and diploid cell substrates for the production of both viral vaccines and other biologicals, such as monoclonal antibodies and products prepared using recombinant DNA technology. These requirements specifically exclude DNA vaccines manufactured in microbial cells. Some of the general manufacturing requirements given here (see sections A.2 and A.3) are also applicable to primary cell substrates. Specific requirements for primary cell cultures can be found in the relevant requirements published by WHO (e.g. production of oral poliomyelitis vaccine in primary monkey kidney cells [25]).

Whenever practicable, manufacturers are encouraged to use cell substrates that can be generated from master cell banks that have been thoroughly characterized.

Requirements published by WHO are intended to be scientific and advisory in nature. The parts of each section printed in normal type have been written in the form of requirements so that, should a national control authority so desire, they may be adopted as they stand as the basis of national requirements. The parts of each section printed in small type are comments or recommendations for guidance.

Part A. General Manufacturing Requirements Applicable to all Types of Cell Culture Production

A.1 Definitions

Cell bank: A cell bank is a collection of ampoules containing material of uniform composition stored under defined conditions, each ampoule containing an aliquot of a single pool of cells.

Cell seed: A quantity of well characterized cells of human, animal or other origin stored frozen at -100°C or below in aliquots of uniform composition derived from a single tissue or cell, one or more of which would be used for the production of a master cell bank.

Master cell bank: A quantity of fully characterized cells of human, animal or other origin stored frozen at -100°C or below in aliquots of uniform composition derived from the cell seed. The master cell bank is itself an aliquot of a single pool of cells generally prepared from a selected cell clone under defined conditions, dispensed into multiple containers and stored under defined conditions. The master cell bank is used to derive all working cell banks. The testing performed on a replacement master cell bank (derived from the same cell clone, or from an existing master or working cell bank) is the same as for the initial master cell bank, unless a justified exception is made.

Working cell bank: A quantity of cells of uniform composition derived from the master cell bank at a finite passage level, dispensed in aliquots into individual containers appropriately stored, usually frozen at -100°C or below, one or more of which would be used for production purposes. All containers are treated identically and, once removed from storage, are not returned to the stock.

Production cell cultures: A collection of cell cultures used for biological production that have been prepared together from one or more containers from the working cell bank or, in the case of primary cell cultures, from the tissues of one or more animals.

Adventitious agents: Contaminating microorganisms of the cell culture or line including bacteria, fungi, mycoplasmas and viruses that have been unintentionally introduced.

In vitro culture age: Duration between the thawing of the master cell bank container(s) and the harvest of the production vessel's cell culture as measured by elapsed chronological time in culture, by the population doubling level of the cells, or by the passage level of the cells when subcultivated by a defined procedure for dilution of the culture.

A.2 Good manufacturing practices

The general manufacturing requirements contained in Good Manufacturing Practices for Pharmaceutical [26] and Biological [27] Products shall apply. Where open manipulations of cells are performed, simultaneous open manipulations of other cell lines shall be avoided to prevent cross-contamination.

Cell cultures shall be prepared by staff who have not, on the same working day, handled animals or infectious microorganisms. The personnel concerned shall be periodically examined medically and found to be healthy.

Particular attention shall be given to the recommendations in Good Manufacturing Practices for Biological Products [27] regarding the training and experience of the staff in charge of production and testing and of those assigned to various areas of responsibility in the manufacturing establishment, as well as to registration of such personnel with the national control authority.

Penicillin or other β-lactam antibiotics shall not be present in production cell cultures.

Minimal concentration of other antibiotics may be acceptable. However, the presence of any antibiotic in a biological process or product is discouraged.

A.2.1 *Selection of source materials*

For all types of cells, the donor shall be free of communicable diseases or diseases of uncertain etiology, such as Creutzfeld-Jakob disease for humans and bovine spongiform encephalopathy (BSE) for cattle.

The national control authority may allow specific exceptions concerning donor health (e.g. myeloma and other tumour cells).

Cells of neurological origin may contain or be capable of amplifying the agent causing spongiform encephalopathies, and shall not be used in the manufacture of medicinal products, apart from cases for which a reasoned exception has been made [28].

The national control authority shall approve source(s) of animal-derived raw materials, such as serum and trypsin. These materials shall comply with the guidelines given in the *Report of a WHO Consultation on Medicinal and other Products in relation to Human and Animal Transmissible Spongiform Encephalopathies* [29]. They shall be subjected to appropriate tests for quality and freedom from contamination by viruses, fungi, bacteria and mycoplasmas to evaluate their acceptability for use in production.

The reduction and elimination from the manufacturing process of raw materials derived from animals and humans is encouraged where feasible.

For some animal-derived raw materials used in the cell culture medium, such as insulin or transferrin, validation of the production process for the elimination of viruses can substitute for virus detection tests.

A.3 Tests applicable to all types of cell cultures

A.3.1 *Tests for viral agents*

Tests shall be undertaken to detect, and where possible identify, any endogenous or exogenous viral agents that may be present in the cells. Special attention shall be given to tests for agents known to be latent in the species from which the cells were derived (e.g. simian virus 40 in rhesus monkeys).

For primary cell cultures, the principles and procedures outlined in Part C, Requirements for Poliomyelitis Vaccine (Oral) [25], together with those in section A.4 of Requirements for Measles, Mumps and Rubella Vaccines and Combined Vaccine (Live) [30] may be followed. For continuous cell lines and diploid cell substrates see parts B and C below.

A.3.2 *Serum used in cell-culture media*

Serum used for the propagation of cells shall be tested to demonstrate freedom from cultivable bacteria, fungi and mycoplasmas, as specified in Part A, sections 5.2 [31] and 5.3 [32] of the revised Requirements for Biological Substances N° 6 (General Requirements for the Sterility of Biological Substances), and from infectious viruses.

Suitable tests for detecting viruses in bovine serum are given in Appendix 1 of the revised Requirements for Biological Substances N° 7 (Requirements for Poliomyelitis Vaccine, Oral) [25]. Where appropriate, more sensitive methods may be used.

In some countries, sera are also examined for freedom from certain phages.

In some countries, irradiation is used to inactivate potential contaminant viruses.

The acceptability of the source(s) of serum of bovine origin shall be approved by the national control authority (see A.2.1).

Human serum shall not be used. If human albumin is used, it shall meet the revised Requirements for Biological Substances N° 27 (Requirements for the Collection, Processing and Quality Control of Blood, Blood Components and Plasma Derivatives) [33], as well as the guidelines contained in *Report of a WHO Consultation on Medicinal and other Products in relation to Human and Animal Transmissible Spongiform Encephalopathies* [29].

A.3.3 *Trypsin used for preparing cell cultures*

Trypsin used for preparing cell cultures shall be tested and found free of cultivable bacteria, fungi, mycoplasmas and infectious viruses, especially bovine or porcine parvoviruses, as appropriate. The methods used to ensure this shall be approved by the national control authority.

The source(s) of trypsin of bovine origin shall be approved by the national control authority (see A.2.1).

In some countries, irradiation is used to inactivate potential contaminant viruses.

A.3.4 *Tests for bacteria, fungi and mycoplasmas at the end of production*

A volume of 20 ml of the pooled supernatant fluids from the production cell cultures shall be tested for bacteria, fungi and mycoplasmas as specified in Part A, sections 5.2 [31] and 5.3 [32] of the Requirements for Biological Substances N° 6 (General Requirements for the Sterility of Biological Substances) by a method approved by the national control authority.

A.3.5 *Tests for adventitious viruses at the end of production*

For virus-based products, control cell cultures are necessary when the product interferes with the test systems used to monitor the absence of adventitious agents. These control cell cultures shall be observed at the end of the production period for viral cytopathic effects and tested for haemadsorbing viruses. If multiple harvest pools are prepared at different times, the cultures shall be observed and tested at the time of the collection of each pool.

In some countries, 25% of the control cell cultures are tested for haemadsorbing viruses using guinea-pig red cells. If the red cells have been stored, the duration of storage should not have exceeded 7 days, and the temperature of storage should have been in the range 2-8°C. In tests for haemadsorbing viruses, calcium and magnesium ions should be absent from the medium.

In some countries, the national control authority also requires that other types of red cells, including cells from humans (blood group IV O), monkeys and chickens (or other avian species), should be used in addition to guinea-pig cells. In all tests, readings should be taken after incubation for 30 minutes at 0-4°C, and again after a further incubation for 30 minutes at 20-25°C. For the test with monkey red cells, readings should also be taken after a final incubation for 30 minutes at 34-37°C.

For recombinant DNA proteins, monoclonal antibodies and other cell-based products, the unprocessed bulk harvest or a lysate of cells and their production culture medium shall be tested.

At the time of production of each unprocessed bulk pool, an appropriate volume of the pool shall be inoculated onto monolayer cultures of at least the following cell types:

- Cultures (primary or continuous cell line) of the same species and tissue type as that used for production. This may not be possible for some continuous cell lines (e.g. hybridomas).
- Cultures of a human diploid cell line.
- Cultures of another cell line from a different species.

The unprocessed bulk-pool sample to be tested shall be diluted as little as possible. Material from at least 10^7 cells and spent culture fluids shall be inoculated onto each of the three cell types. The resulting co-cultivated cell cultures shall be observed for evidence of adventitious viruses for at least 2 weeks. If the product is from a continuous cell line known to be capable of supporting the growth of human cytomegalovirus, human diploid cell cultures shall be observed for at least 4 weeks.

Extended cell culture for the purposes of identifying human cytomegalovirus can be replaced by the use of specific probes to detect cytomegalovirus nucleic acid.

At the end of the observation period, aliquots of each of the three co-cultivated cell culture systems shall be tested for haemadsorbing viruses.

PART B. REQUIREMENTS FOR CONTINUOUS-CELL-LINE SUBSTRATES

B.1 General considerations

Several types of continuous cell line have been employed as substrates in the production of biologicals, including Vero cells in the preparation of live and inactivated viral vaccines and the use of CHO cells in the production of a number of recombinant proteins. The advantage of such cell lines is that they grow relatively rapidly and provide high yields of monolayer or, in some cases, suspension cultures.

Continuous cell lines may have biochemical, biological and genetic characteristics that differ from primary or diploid cells. In particular, they may produce transforming proteins and may contain potentially oncogenic DNA. In some cases, continuous cell lines may cause tumours when inoculated into animals. The manufacturing process for the production of biologicals in continuous-cell-line substrates should take these factors into account in order to ensure the safety of the product. Generally, purification procedures will result in the extensive removal of cellular DNA, other cellular components and potential adventitious agents. Procedures that extensively degrade or denature DNA might be appropriate for some products (e.g. rabies vaccine). When continuous cell lines are being contemplated for use in the development of live viral vaccines, careful consideration must be given to the possible incorporation of oncogenic cellular DNA into the virions.

Production of biologicals from continuous-cell-line substrates should be based on well defined master and working cell banks. The master cell bank is generally derived from a selected cell clone. The working cell bank is derived by expansion of one or more containers of the master cell bank.

Evidence that the cell line is free from cultivable bacteria, mycoplasmas, fungi and infectious viruses, and where appropriate, potentially oncogenic adventitious agents should be provided. Special attention should be given to viruses that commonly contaminate the animal species from which the cell line is derived. Cell seed should preferably be free from all adventitious agents. However, certain cell lines express endogenous viruses, e.g. retroviruses. Tests capable of detecting such agents should be carried out on cells grown under production conditions, and the results should be reported. Specific contaminants identified as endogenous agents in the master and working cell banks should be shown to be inactivated and/or removed by the purification procedure used in production. The validation of the purification procedure used is also considered essential [34] (see Appendix).

The data required for the characterization of any continuous cell line to be used for the production of biologicals include: a history of the cell line and a detailed description of the production of the cell banks, including methods and reagents used during culture, in vitro culture age, and storage conditions; the results of tests for infectious agents; distinguishing features of the cells, such as biochemical, immunological or cytogenetic patterns which allow them to be clearly distinguished from other cell lines; and the results of tests for tumorigenicity, including data from the scientific literature.

Special consideration should be given to products derived from cells that contain known viral genomes (e.g.Namalva cells). Cells modified by recombinant DNA technology have been increasingly used in the manufacture of novel medicinal products and specific considerations for those products are addressed elsewhere [35, 36].

Continuous cell lines should be characterized so that appropriate controls for the purity and safety of the final product can be included. For example, if a continuous cell line contains an endogenous virus, tests to ensure the absence of any detectable biological activity of that virus could be incorporated as one of the requirements for products derived from that cell line. Alternatively, process validation may replace testing at the end of production for endogenous viruses when a high degree of assurance of consistency of virus clearance can be provided.

There has been considerable discussion internationally on general criteria for the acceptability of products (e.g. hormones, blood components, viral vaccines) prepared from continuous cell lines. A consensus has emerged on the general desirability of achieving a high degree of purification of the product, involving significant removal or destruction of DNA of cell substrate origin. Manufacturers considering the use of continuous cell lines should be aware of the need to develop and evaluate efficient methods for purification as an essential element of any product development programme.

While all continuous cell lines, by definition, have an infinite life span, they may express no tumorigenic properties below a certain passage (or population-doubling) level, but subsequently display increasing evidence of the tumorigenic phenotype with increasing passage. It is therefore important to establish an age limit for in vitro cultures beyond which they cannot be used for production. The limit should be based on data derived from production cells expanded under pilot plant-scale or full-scale conditions to the proposed in vitro culture age limit or beyond. Generally, the production cells are obtained by expansion of the working cell bank; however, the master cell bank could be used to prepare the production cells, given appropriate justification. Increases in the established in vitro culture age limit for production should be supported by data from cells that have been expanded to an in vitro culture age that is equal to or greater than the proposed new limit.

The following Requirements concern the characterization and testing of continuous cell lines used for the production of biologicals. They should be read in conjunction with the general manufacturing requirements applicable to all cell cultures contained in part A of these Requirements. Specific requirements for purity as well as other quality control procedures will be incorporated in requirements published by WHO for individual biological products.

B.2 Manufacturing requirements

B.2.1 *Certification of continuous cell lines for use in the production of biologicals*

A continuous cell line line used for biologicals production shall be approved by the national control authority and shall be identified by historical records that include information on the origin of the cell line, its method of development and the in vitro culture age limit for production.

A continuous cell line used for bilogicals production shall also be characterized with respect to genealogy, genetic markers (e.g. histo-compatibility leukocyte antigen (HLA), DNA fingerprinting), viability during storage, and growth characteristics at passage levels (or population doublings or time-in-culture, as appropriate) equivalent to, or beyond, those of the master and working cell banks and the cell cultures used for production.

B.2.2 *Cells banks*

The use of continuous cell lines for the manufacture of biological products shall be based on the cell bank system, which shall include a well defined master cell bank and working cell bank.

The cell bank used for the production of biologicals shall be that approved by and registered with the national control authority. The continuous cell line from which the master cell bank has been derived shall be characterized as described in section B.1. The working cell bank shall be shown to yield cell cultures capable of producing biologicals that are both safe and efficacious in humans.

In section B.2.3, extensive testing directed at identifying exogenous and endogenous agents that may be present in the cell line is described; special attention is given to agents known to be present in a latent state in the species from which the cells were derived. Such extensive testing need only be performed once, on either the master cell bank or a working cell bank. Once a continuous cell line has been characterized in this respect, further testing of working cell banks or production cell cultures is restricted to tests directed at detecting common adventitious agents that could have contaminated the cultures during their preparation.

The tumorigenicity testing described in section B.2.3.7 shall be performed only once on cells of either the master cell bank or a working cell bank propagated to an in vitro culture age at or beyond the limit for production. If the cell line has already been documented to be tumorigenic or if the class of cells to which it belongs (e.g. hybridomas) is tumorigenic, the cell line may be presumed to be tumorigenic and tumorigenicity tests need not be undertaken.

Both the master and working cell banks shall be stored at -100°C or below (i.e. in either the liquid or vapour phase of liquid nitrogen). The location, identity and inventory of individual ampoules of cells shall be thoroughly documented.

It is recommended that the master and working cell banks should each be stored in at least two widely separated areas within the production facility in order to avoid accidental loss of the cell line.

B.2.3 *Identification and characteristics of continuous cell lines*

The characterization of a continuous cell line intended for use in the manufacture of biologicals shall include information on: the history and general characteristics of the cell line; the cell bank system; and quality control testing. These data shall be made available to the national control authority.

B.2.3.1 *Identity test*

The cell banks shall be identified by a method approved by the national control authority.

Methods for identity testing include, but are not limited to, biochemical (e.g. isoenzyme analyses), immunological (e.g. HLA assays), cytogenetic tests (e.g. for chromosomal markers), and tests for genetic markers (DNA finger-printing).

B.2.3.2 *Sterility tests*

A volume of 20 ml of supernatant fluids from cell cultures derived from at least one ampoule of the master and working cell banks shall be tested for bacteria, fungi and mycoplasmas. Tests shall be performed as specified in Part A, sections 5.2 [31] and 5.3 [32] of the revised Requirements for Biological Substances N° 6 (General Requirements for the Sterility of Biological Substances), by a method approved by the national control authority.

B.2.3.3 *Tests for viral agents using cell cultures*

Live or disrupted cells and spent culture fluids of the master or working cell bank shall be inoculated onto monolayer cultures or co-cultivated with monolayer cultures, as appropriate, of the following cell types:

- Cultures (primary or continuous cell line) of the same species and tissue type as the continuous cell line. This may not be possible for some continuous cell lines, e.g. hybridomas.
- Cultures of a human diploid cell line.
- Cultures of another cell line from a different species.

The sample to be tested shall be diluted as little as possible. Material from at least 10^7 cells and spent culture fluids shall be inoculated onto each of the three cell types. The resulting cultures shall be observed for at least 2 weeks for evidence of adventitious viruses. If the continuous cell line being tested is known to be capable of supporting the growth of human cytomegalovirus, human diploid cell cultures shall be observed for at least 4 weeks.

Extended cell culture for the purposes of identifying human cytomegalovirus can be replaced by the use of specific probes to detect cytomegalovirus nucleic acid.

At the end of the observation period, aliquots of each of the three cell culture systems shall be tested for haemadsorbing viruses.

B.2.3.4 *Tests for viral agents using animals and eggs*

The cells of the master and working cell banks are suitable for production if none of the animals or eggs shows evidence of the presence of any viral agent attributable to the cell banks.

Tests in animals: Tests in animals for pathogenic viruses shall include the inoculation by the intramuscular route of each of the following groups of animals with cells from the master or working cell banks propagated to or beyond the maximum in vitro culture age (or population doubling, as appropriate) used for production, where at least 10^7 viable cells are divided equally among the animals in each group:

- Two litters of suckling mice, comprising a total of at least ten animals, less than 24 h old; and
- Ten adult mice weighing 15-20 g.

In some circumstances, tests in five guinea-pigs weighing 350-450 and five rabbits weighing 1.5-2.5 kg may be considered.

The test in rabbits for the presence of B virus in cell lines of simian origin may be replaced by a test in rabbit kidney-cell cultures.

The animals shall be observed for at least 4 weeks. Any animals that are sick or show any abnormality shall be investigated to establish the cause. The test is not valid if more than 20% of the animals in the test group become sick for non-specific reasons and do not survive the observation period.

In some countries, the suckling and adult mice are also inoculated by the intracerebral route.

If the cell line is of rodent origin, at least 10^6 viable cells shall be injected intracerebrally into each of ten susceptible adult mice to test for the presence of lymphocytic choriomeningitis virus.

Tests in eggs: At least 10^6 viable cells from the master or working cell banks, propagated to or beyond the maximum in vitro culture age (or population doubling, as appropriate) shall be injected into the allantoic cavity of each of ten embryonated chicken eggs, and the yolk sac of each of another ten embryonated chicken eggs. The eggs shall be examined after not less than 5 days of incubation. The allantoic fluids of the eggs shall be tested with red cells from guinea-pig and chickens (or other avian species) for the presence of haemagglutinins. The test is not valid if more than 20% of the embryonated chicken eggs in the test group are discarded for non-specific reasons.

Usually, the eggs used for the yolk sac test should be 5-6 days old. The eggs used for the allantoic cavity test should be 9-11 days old.

Alternative ages for the embryonated chicken eggs and alternative incubation periods are acceptable if they have been determined to be capable of detecting the presence of routine adventitious agents in the test samples.

B.2.3.5 *Tests for retroviruses and other endogenous viruses or viral nucleic acid*

Test samples from the master or working cell banks, propagated to or beyond the maximum in vitro culture age (or population doubling, as appropriate) shall be examined for the presence of retroviruses using the following techniques:

- Infectivity assays (if the infectivity assay is positive, tests for reverse transcriptase are not necessary);
- Transmission electron microscopy (TEM); and
- reverse transcriptase (RTase) assays (performed in the presence of magnesium and manganese) on pellets obtained from fluids by high speed centrifugation (e.g. 125 000 g for 1 h) at 4°C.

Recently developed highly sensitive RTase assays may be considered, but the results need to be interpreted with caution because RTase activity is not unique to retroviruses and may derive from other sources, such as retrovirus-like elements which do not encode a complete genome or cellular DNA polymerase.

It is often possible to increase the sensitivity of cell-culture infectivity assays by first inoculating the test material onto human cell lines that can support retroviral growth in order to amplify any retrovirus contaminant that may be present at low concentrations. For non-murine retroviruses, test cell lines should be selected for their capacity to support the growth of a broad range of retroviruses, including viruses of human and non-human primate origin [37, 38].

For murine retroviruses, amplification of low-level contaminants may be achieved by co-cultivation of cells with a highly susceptible cell line, e.g. *Mus dunni* cells [39]. The latter are susceptible to infection by all tested murine leukaemia viruses except Moloney murine leukaemia virus. For that reason, another susceptible cell, for example SC-1 [40], should also be used. Fluid from the resulting co-cultures should be further passaged on *Mus dunni* or other susceptible cells and subsequently assayed for murine leukaemia virus.

A variety of other assays may be useful, depending on the circumstances. Some examples of such assays include viable cell immunofluorescence (IFA) on *Mus dunni* cells co-cultivated with the test cells using a broadly reactive monoclonal antibody (e.g. HY95) for the detection of ecotropic, xenotropic, mink-cell focus-forming and amphotropic viruses; feline S + L assays using PG4 cells [41] for detection of amphotropic viruses; mink S + L assays for detection of xenotropic viruses [12] and mouse S + L assays using D56 [42] cells for detection of ecotropic viruses.

Murine and other rodent cell lines or hybrid cell lines containing a rodent component should be assumed to be inherently capable of producing infectious retroviruses. For murine cell lines used for monoclonal antibody production, the extent of testing for specific retroviruses may be reduced. However, the manufacturing process should be evaluated for removal and/or inactivation of retroviruses. For murine-human hybrid cell lines, additional concerns arise. Any proposed testing should be discussed with the national control authority on a case-by-case basis.

Probe hybridization/polymerase-chain-reaction amplification and virus-specific monoclonal antibody detection may provide additional information on the presence or absence of specific contaminants.

B.2.3.6 *Tests for selected viruses*

The following tests shall be undertaken on a selected basis on samples from the working cell bank propagated to or beyond the maximum in vitro culture age (or population doubling, as appropriate).

Murine cell lines shall be tested for species-specific viruses using mouse, rat and hamster antibody production tests. In vivo testing for lymphocytic choriomeningitis virus, including a challenge for non-lethal strains, is required for such cell lines as specified in B.2.3.4.

Human cell lines shall be screened for human viral pathogens such as Epstein-Barr virus, cytomegalovirus, human retroviruses, and hepatitis B and C viruses with appropriate in vitro techniques. Selection of the viruses to be screened for shall take into account the tissue source and medical history of the person from whom the cell line was derived. Tests for retroviruses are specified in section B.2.3.5.

The use of other cell cultures also may be appropriate for the characterization of cell banks, depending on the cell type and source of the cell line being characterized [17]. Under certain circumstances, specific testing for the presence of other transforming viruses, such as papillomavirus, adenovirus and herpesvirus 6 and 7, may also be indicated.

B.2.3.7 *Tests for tumorigenicity*

If the continuous cell line has already been demonstrated to be tumorigenic (e.g. BHK21, CHO, C127), or if the class of cells to which it belongs, for example hybridoma, is tumorigenic, it is not necessary to require additional tumorigenicity tests. A new cell line shall be presumed to be tumorigenic unless data demonstrate that it is not. If a manufacturer proposes to characterize the cell line as non tumorigenic, the following tests shall be undertaken.

Tests in vivo: Cells from the master or working cell bank propagated to or beyond the in vitro culture age limit for production shall be examined for tumorigenicity in a test approved by the national control authority. The test shall involve a comparison between the continuous cell line and a suitable positive reference preparation (e.g. HeLa, Hep 2 or FL cells).

A negative control is not essential but desirable. For that purpose non-tumorigenic diploid cell lines such as WI-38 or MRC-5 may be used.

Animal systems that have been shown to be suitable for this test include:

(a) Athymic mice (*Nu/Nu* genotype);

(b) Newborn mice, rats or hamsters that have been treated with antithymocyte serum or globulin; and

(c) Thymectomized and irradiated mice that have been reconstituted (T-, B+) with bone marrow from healthy mice.

Whichever animal system is selected, the cell line and the reference cells are injected into separate groups of ten animals each. In both cases, the inoculum for each animals is 10^7 cells suspended in a volume of 0.2 ml, and the injection is by either the intramuscular or the subcutaneous route. In the case of newborn animals (b), the animals are treated with 0.1 ml of antithymocyte serum or globulin on days 0, 2, 7 and 14 after birth. A potent serum or globulin is one that supresses the immune mechanisms of the growing animals to the extent that the subsequent inoculum of 10^7 positive reference cells regularly produces tumours and metastases.

At the end of the observation period all animals, including the reference group(s), shall be killed and examined for gross and microscopic evidence of the proliferation of inoculated cells at the site of injection and in other organs (e.g. lymph nodes, lungs, kidneys and liver).

In all test systems, the animals shall be observed and palpated at regular intervals for the formation of nodules at the sites of injection. Any nodules formed should be measured in two perpendicular dimensions, the measurements being recorded regularly to determine whether there is progressive growth of the nodule. Animals showing nodules which begin to regress during the period of observation shall be killed before the nodules are no longer palpable, and processed for histological examination. Animals with progressively growing nodules shall be observed for 1-2 weeks. Among those without nodule formation, half shall be observed for 3 weeks and half for 12 weeks before they are killed and processed for histological examination. A necropsy shall be performed on each animal and shall include examination for gross evidence of tumour formation at the site of inoculation and in other organs such as lymph nodes, lungs, brain, spleen, kidneys and liver. All tumour-like lesions and the site of inoculation shall be examined histologically. In addition, since some cell lines may give rise to metastases without evidence of local tumour growth, any detectable regional lymph nodes and the lungs of all animals shall be examined histologically.

For the test to be considered valid, progressively growing tumours must be produced in at least nine of ten animals injected with the positive reference cells.

In vitro tests may be considered sufficient by some national control authorities.

Two in vitro tests that have been found to provide useful additional information on tumorigenicity are: colony formation in soft agar gels, and production of invasive cell growth following inoculation onto organ cultures. They may be used to characterize more fully the cell lines that show no evidence of tumorigenicity in animal tests (see above), or when the results are equivocal.

As the cells used in the production of biologicals may contain activated oncogenes, assays of cell transformation with DNA derived from a continuous cell line at the limit for in vitro culture age for production should be considered in order to determine whether or not activated oncogenes can be detected. The 3T3 assay system has been found useful for *ras* assays. Additonal tests may also be considered as new techniques are developed for the detection of a broader range of oncogenes.

B.2.3.8 *Tests on cells carrying a recombinant-DNA expression system*

Data shall be obtained demonstrating that a continuous cell line can be used for its intended purpose. If a continuous cell line contains an expression construct to produce a recombinant DNA-derived protein, data shall be obtained to demonstrate the consistent quality and quantity of the protein it produces throughout the proposed in vitro culture age range for production [14, 15]. Studies shall be performed to determine whether manipulation of the cell line in order to produce a product by transfection changes its biological chacteristics significantly, for instance conversion to the tumorigenic phenotype. Any such change must be taken into account in product development and in assessing approaches taken to assure an acceptable product.

The International Conference on Harmonisation has issued additional useful information [43].

B.2.4 *Production cell cultures*

Characterization of the product and routine monitoring for adventitious agents during the production process are part of the quality control of biological products.

The choice of method for quality control of the production cell substrate depends on the nature of the propagation system used. Cell substrates are propagated as monolayer cultures, in suspension cultures or in bioreactors, and can be maintained on a short-term, a long-term or even on a potentially indefinite basis. The product is obtained either from a single harvest of cell culture fluid or from multiple harvests. In some cases, quality control testing may need to be performed on each harvest before pooling into a bulk lot. The management of cell substrates for the purposes of quality control testing should be designed to optimize sensitivity of the testing.

B.2.4.1 *Serum used in cell-culture media*

Serum used in cell-culture media shall be tested as specified in section A.3.2.

B.2.4.2 *Trypsin used for preparing cell cultures*

Trypsin used for preparing cell cultures shall be tested as specified in section A.3.3.

B.2.4.3 *Identity test*

For viral vaccines, an identity test shall be performed on the control cell culture as described in section B.2.3.1. For recombinant DNA proteins and monoclonal antibodies, the presence of the protein at consistent levels in the harvest is an adequate confirmation of identity and purity.

B.2.4.4 *Tests for bacteria, fungi and mycoplasmas at the end of production*

Tests for bacteria, fungi and mycoplasmas shall be conducted on the production culture supernatant or lysate as specified in section A.3.4.

B.2.4.5 *Tests for adventitious viruses at the end of production*

Tests for adventitious viruses shall be conducted on the production culture supernatant or lysate as specified in section A.3.5.

Part C. Requirements for Diploid Cell Substrates

C.1 General considerations

Two human diploid cell lines, WI-38 and MRC-5, derived from embryo lung tissue, have been in widespread use for many years for the production of live virus vaccines, including oral poliomyelitis, measles, mumps, rubella and varicella vaccines, and inactivated vaccines, for example, rabies and hepatitis A vaccines. In addition, a rhesus diploid cell line, FRhL-2, has been in limited use for rabies vaccine production. These substrates have been found to be safe and to produce vaccines that stimulate effective immunity without untoward reactions attribute to the cell substrate.

The following requirements concern the characterization and testing of diploid cell lines used for the production of biologicals. They should be read in conjunction with the general manufacturing requirements applicable to all cell cultures contained in Part A of these Requirements.

C.2 Manufacturing requirements

C.2.1 *Certification of diploid cell lines for use in the production of biologicals*

A diploid cell line used for biologicals production shall be approved by the national control authority and shall be identified by historical records that include information on the origin of the cell line, its method of development and the range of passage levels at which it can used in biologicals production.

A new diploid cell line (e.g. other than WI-38, MRC-5 and FRhL-2) used for biologicals production shall be characterized with respect to genealogy, genetic markers (e.g. HLA, DNA fingerprinting), or other markers of identity acceptable to the national control authority, as well as for viability during storage. In addition, data must be obtained to establish the cell line's diploid character and growth characteristics at in vitro culture ages equivalent to, or beyond, those of the master and working cell banks, and of the cell cultures used for production.

Accumulated experience suggests that WI-38 and MRC-5 can be used for production until 10 generations before senescence.

C.2.2 *Cell banks*

C.2.2.1 *Master cell bank and working cell bank*

Tests shall be performed on the master and working cell banks as described in section C.2.3, where appropriate and approved by the national control authority. In addition, for a new diploid cell line (e.g. other than WI-38, MRC-5, and FRhL-2) the cells of the working cell bank shall be shown to be diploid and stable with respect to karyology by the tests outlined in section C.2.3.5.

C.2.3 *Identification and characteristics of diploid cell lines*

The characterization of a diploid cell intended for use in the manufacture of biologicals shall include information on: the history and general characteristics of the cell line; the cell bank system; and quality control testing. These data shall be made available to the national control authority.

C.2.3.1 *Identity tests*

An identity test shall be performed on the master cell bank by a method approved by the national control authority.

Methods for identity testing include, but are not limited to, biochemical tests (e.g. isoenzyme analyses), immunological tests (e.g. HLA assays), cytogenetic tests (e.g. for chromosomal markers), and tests for genetic markers (DNA fingerprinting).

Tests to ensure that the master cell bank is not contaminated with a continuous cell line shall be performed.

Tests of identity such as DNA fingerprinting of appropriate sensitivity, karyology at different levels of passage or studies of lifespan in culture may be used for this purpose if approved by the national control authority.

C.2.3.2 *Sterility tests*

Tests for bacteria, fungi and mycoplasmas shall be conducted in cell cultures as specified in section B.2.3.2.

C.2.3.3 *Tests for viral agents using cell cultures*

Tests for viral agents shall be conducted in cell cultures as specified in section B.2.3.3.

C.2.3.4 *Tests for viral agents using animals and eggs*

Test for viral agents shall be conducted in animals and eggs as specified in section B.2.3.4.

C.2.3.5 *Chromosomal characterization of a diploid cell line*

The usefulness of chromosomal characterization depends on the nature of the product and the manufacturing process. In general, products that might contain live cells or which have little "down-stream" purification will require chromosomal characterization of the cell line. Such products manufactured in cells identified to be WI-38, MRC-5 or FRhL-2 cells do not require recharacterization of the cell substrate by karyology, unless the cells have been genetically modified.

The utility of chromosomal monitoring of the cell substrate for unpurified products manufactured in other cell lines shall be evaluated on a case-by-case basis. However, products that contain no cells and are highly purified will not require this test.

For the determination of the general character of a new diploid cell line (i.e. other than WI-38, MRC-5 and FRhL-2), samples from the master cell bank shall be examined at approximately four equally spaced intervals over the life span of the cell line during serial cultivation through to senescence. Each sample shall consist of a minimum of 200 cells in metaphase and shall be examined for exact counts of chromosomes and for frequency of hyperdiploidy, hypoploidy, polyploidy, breaks and structural abnormalities. The acceptability of any new diploid cell line shall be determined by the national control authority.

It is recommended that photographic reconstruction should be employed to prepare chromosome-banded karyotypes of an additional ten metaphase cells.

Stained slide preparations of the chromosomal characterization of the diploid cell line, or photographs of these, shall be maintained permanently as part of the cell line record.

C.2.3.7 *Tests for tumorigenicity*

The tumorigenic potential of a new diploid cell line (i.e. other than WI-38, MRC-5 and FRhL-2) shall be tested as specified in section B.2.3.7 as part of the characterization of the cell line, but is not required on a routine basis.

If satisfactory data from least two independent laboratories are available, further tumorigenicity testing may not be required. The adequacy of tumorigenicity testing of a new diploid cell line should be discussed with the national control authority. Positive results should be discussed with the authority, taking into consideration the purity of the product, including residual cellular DNA.

C.2.3.8 *Tests on cells carrying a recombinant-DNA expression construct*

Data shall be obtained demonstrating that a diploid cell line can be used for its intended purpose. If a cell line contains an expression construct to produce a recombinant-DNA-derived protein, data shall be obtained to demonstrate the consistent quality and quantity of the protein produced throughout the proposed in vitro culture age range for production [33, 34].

The International Conference on Harmonisation has issued additional useful information [43].

C.2.4 ***Production cell cultures***

C.2.4.1 *Serum used in cell-culture media*

Serum used in cell-culture media shall be tested as specified in section A.3.2.

C.2.4.2 *Trypsin used for preparing cell cultures*

Trypsin used for preparing cell cultures shall be tested as specified in section A.3.3.

C.2.4.3 *Identity test*

An identity test shall be performed on the control cell culture as specified in section B.2.3.1.

C.2.4.4 *Tests for bacteria, fungi and mycoplasmas at the end of production*

Tests for bacteria, fungi and mycoplasmas shall be conducted on the production culture supernatant or lysate as specified in section A.3.4.

C.2.4.5 *Tests for adventitious viruses at the end of production*

Tests shall be conducted on the product at the end of production but before further processing as specified in section A.3.5. If the presence of the product interferes, tests shall be performed on the control cell culture as specified in section A.3.5.

Authors

The first draft of these Requirements was prepared by Dr V. Grachev, Scientist, Biologicals, Dr D. Magrath, Chief (1987-1994), Biologicals, and Dr E. Griffiths, Chief (from 1994), Biologicals, World Health Organization, Geneva, Switzerland.

A revised draft was formulated by Dr V. Grachev, Deputy Director, Institute of Poliomyelitis and Viral Encephalitides, Russian Academy of Medical Sciences, Moscow, Russian Federation, Dr E. Griffiths, Chief, Biologicals, WHO, Geneva, Switzerland and Dr J.C. Petricciani, Vice-President, Genetics Institute, Cambridge, MA, USA.

After extensive consultation, including discussion of the main scientific issues at a WHO/IABS meeting held at the Mérieux Foundation, Annecy, France, from 29 September to 1 October 1996, further amendments to the text were proposed by the following group of experts:

Dr Y.Y. Chiu, Director, Division of New Drug Chemistry, Food and Drug Administration, Rockville, MD, USA.

Dr R. Dobbelaer, Acting Chief, Biological Standardization, Institute for Hygiene and Epidemiology, Brussels, Belgium.

Dr V. Grachev, Institute of Poliomyelitis and Viral Encephalitides, Russian Academy of Medical Sciences, Moscow, Russian Federation.

Dr E. Griffiths, Chief, Biologicals, World Health Organization, Geneva, Switzerland.

Dr I. Gust, CSL Limited, Parkville, Victoria, Australia.

Dr M.C. Hardegree, Director, Office of Vaccine Research and Review, Center for Biologics Evaluation and Research, Food and Drug Administration, Bethesda, MD, USA.

Dr T. Hayakawa, National Institute of Health Science, Tokyo, Japan.

Professor F. Horaud, Pasteur Institute, Paris, France.

Dr A.S. Lubiniecki, SmithKline Beecham, King of Prussia, PA, USA.

Dr P. Minor, Head, Division of Virology, National Institute for Biological Standards and Control, Potters Bar, England.

Dr B. Montagnon, Pasteur Mérieux Sera and Vaccines, Marcy l'Etoile, France.

Dr J. Peetermans, SmithKline Beecham Biologicals, Rixensart, Belgium.

Dr J. Petricciani, Vice-President, Genetics Institute, Cambridge, MA, USA.

Dr A. Ridgeway, Head, Biotechnology Section, Health and Welfare Canada, Ottawa, Ontario, Canada.

Dr J. Robertson, National Institute for Biological Standards and Control, Potters Bar, England.

Dr G. Schild, Director, National Institute for Biological Standards and Control, Potters Bar, England.

Dr K.B. Seamon, Immunex Corporation, Seattle, WA, USA.

Acknowledgements

Acknowledgements are due to the following experts for their comments and advice: Dr S.C. Arya, Centre for Logistical Research and Innovation, New Delhi, India; Dr T. Bektimirov, Deputy Director, Tarasevic State Institute for the Standardization and Control of Medical Biological Preparations, Moscow, Russian Federation; Dr M. Duchêne, Director, Quality Control, SmithKline Beecham Biologicals, Rixensart, Belgium; Dr B.D. Garfinkle, Vice-President, Vaccine Quality

Operations, Merck & Co., West Point, PA, USA; Dr R.K. Gupta, Assistant Director, Massachusetts Public Health Biologics Laboratory, Boston, MA, USA; Dr K. Healy, Head, Quality Assurance Department, CSL Limited, Parkville, Victoria, Australia; Dr A. Homma, Regional Adviser in Biologics, WHO Regional Office for the Americas, Washington, DC, USA; Dr J.G. Kreeftenberg, Head, Quality and Regulatory Affairs, National Institute of Public Health and Environmental Protection, Bilthoven, Netherlands; Dr L. Lavese, Head of International Affairs Department, Farmindustria, Rome, Italy; Dr R. Netter, Rue Vaugirard, Paris, France; Dr A.S. Outschoorn, Scientist Emeritus, United States Pharmacopeia, Rockville, MD, USA; Dr E. Walker, Head, Molecular Biology Section, Therapeutic Goods Administration, Woden, Australia.

References

1 Van Wezel AL: Microcarrier technology – present status and prospects, in Griffiths B, Horaud F, Spier R, Hennessen W (eds): Fifth General Meeting of ESACT. Dev Biol Stand. Basel, Karger, 1982, vol 55, pp 3-9.

2 Jacobs JP et al: Guidelines for the acceptability, management and testing of serially propagated human diploid cells for the production of live virus vaccines for use in man. J Biol Stand 1981;9:331-342.

3 Hayflick L, Plotkin S, Stevenson R: History of the acceptance of human diploid cell strains as substrates for human virus vaccine production, in Petricciani JC, Hennessen W (eds): Cells, Products, Safety. Dev Biol Stand, 1987, vol 68, pp 9-17.

4 Grachev V: World Health Organization attitude concerning the use of continuous cell lines as substrate for production of human virus vaccines, in Mizrahi A (ed): Advances in Biotechnological Processes Viral Vaccines. Wiley-Liss, 1990, vol 14, pp 37-67.

5 Acceptability of cell substrates for production of biologicals. Report of a WHO Study Group. Geneva, World Health Organization, 1987 (WHO Technical Report Series, N° 747).

6 Requirements for continuous cell lines used for biologicals production, in WHO Expert Committee on Biological Standardization. Thirty-sixth Report. Geneva, World Health Organization, 1987, Annex 3 (WHO Technical Report Series, N° 745).

7 WHO bank of Vero cells for the production of biologicals, in WHO Expert Committee on Biological Standardization. Fortieth Report. Geneva, World Health Organization, 1990, p 11 (WHO Technical Report Series, N° 800).

8 Temin HM: Overview of biological effects of addition of DNA molecules to cells. Journal of Medical Virology 1990;31:13-17.

9 Kurth R: Risk potential of the chromosomal insertion of foreign DNA. Annals of the New York Academy of Sciences 1995;772:140-150.

10 Petricciani JC, Horaud FN: DNA, dragons, and sanity. Biologicals 1995;23:233-238.

11 Nichols WW et al: Potential DNA vaccine integration into host cell genome. Annals of the New York Academy of Sciences 1995;772:30-38.

12 Coffin JM: Molecular mechanisms of nucleic acid integration. Journal of Medical Virology 1990;31:43-49.

13 Petricciani JC, Regan PJ: Risk of neoplastic transformation from cellular DNA: calculations using the oncogene model. Dev Biol Stand 1986;68:43-49.

14 Strain AJ: The uptake and fate of DNA transfected into mammalian cells in vitro. Dev Biol Stand 1986;68:27-32.

15 Doehmer J: Residual cellular DNA as a potential transforming factor. Dev Biol Stand 1986; 68:33-41.

16 Wierenga DE, Cogan J, Petricciani JC: Administration of tumor cell chromatin to immunosuppressed and non-immunosuppressed primates. Biologicals 1995;23:221-224.

17 Greenwald P et al: Morbidity and mortality among recipients of blood from pre-leukemic and pre-lymphomatous donors. Cancer 1976;38:324-328.

18 Fleckenstein B, Daniel MD, Hunt RD: Tumor inductions with DNA of oncogenic primate herpesviruses. Nature 1978;274:57-59.

19 Duxbury M et al: DNA in plasma of human blood for transfusion. Biologicals 1995;23:229.

20 Nawroz H et al: Microsatellite alterations in serum DNA of head and neck cancer patients. Nature Medicine 1996;2:1035-1037.

21 Perrin P, Morgeaux S: Inactivation of DNA by beta-propiolactone. *Biologicals* 1995;23:207.

22 Requirements for Poliomyelitis Vaccine (Inactivated), in Requirements for Biological Substances: 1. General Requirements for Manufacturing Establishments and Control Laboratories; 2. Requirements for Poliomyelitis Vaccine (Inactivated). Report of a Study Group. Geneva, World Health Organization, 1959 (WHO Technical Report Series, N° 178).

23 Requirements for Poliomyelitis Vaccine (Inactivated), in Requirements for Biological Substances. Manufacturing and Control Laboratories. Report of a WHO Expert Group. Geneva, World Health Organization, 1966 (WHO Technical Report Series, N° 323).

24 Hilleman M: Cells, vaccines and pursuit of precedent: Bethesda, MD, National Cancer Institute, 1968 (National Cancer Institute Monograph, N° 29).

25 Requirements for Poliomyelitis Vaccine (Oral) (Requirements for Biological Substances N° 7, revised 1989), in WHO Expert Committee on Biological Standardization. Fortieth Report. Geneva, World Health Organization, 1990, Annex 1 (WHO Technical Report Series, N° 800).

26 Good manufacturing practices for pharmaceutical products, in WHO Expert Committee on Specifications for Pharmaceutical Preparations. Thirty-second Report. Geneva, World Health Organization, 1992, Annex 1 (WHO Technical Report Series, N° 823).

27 Good manufacturing practices for biological products, in WHO Expert Committee on Biological Standardization. Forty-second Report. Geneva, World Health Organization, 1992, Annex 1 (WHO Technical Report Series, N° 822).

28 Questions de santé publique liées aux encéphalopathies spongiformes chez l'animal et chez l'homme: Mémorandum d'une réunion de l'OMS. Bulletin of the World Health Organization 1992;70:573-582.

29 Report of a WHO Consultation on Medicinal and Other Products in Relation to Human and Animal Transmissible Spongiform Encephalopathies. Geneva, World Health Organization, 1997 (unpublished document WHO/BLG/97.2; available on request from Biologicals, World Health Organization, 1211 Geneva 27, Switzerland).

30 Requirements for Measles, Mumps and Rubella Vaccines and Combined Vaccine (Live) (Requirements for Biological Substances N° 47, 1992), in WHO Expert Committee on Biological Standardization. Forty-fourth Report. Geneva, World Health Organization, 1994, Annex 3 (WHO Technical Report Series, N° 840).

31 General Requirements for the Sterility of Biological Substances (Requirements for Biological Substances N° 6, revised 1973), in WHO Expert Committee on Biological Standardization. Twenty-fifth Report. Geneva, World Health Organization, 1973, Annex 4 (WHO Technical Report Series, N° 530).

32 General Requirements for the Sterility of Biological Substances (Requirements for Biological Substances N°6, revised 1973, amendment 1995), in WHO Expert Committee on Biological Standardization. Forty-sixth Report. Geneva, World Health Organization, 1988, Annex 3 (WHO Technical Report Series, N° 872).

33 Requirements for the Collection, Processing and Quality Control of Blood, Blood Components and Plasma Derivatives (Requirements for Biological Substances N° 27, revised 1992), in WHO Expert Committee on Biological Standardization. Forty-third Report. Geneva, World Health Organization, 1994, Annex 2 (WHO Technical Report Series, N° 840).

34 Viral safety evaluation of biotechnology products. Geneva, International Conference on Harmonisation, 1997 (unpublished document available on request from International Conference on Harmonisation Secretariat, International Federation of Pharmaceutical Manufacturers Associations, 30 rue de St-Jean, 1211 Geneva 18, Switzerland).

35 Guidelines for assuring the quality of pharmaceutical and biological products prepared by recombinant DNA technology, in WHO Expert Committee on Biological Standardization. Forty-first Report. Geneva, World Health Organization, 1991, Annex 3 (WHO Technical Report Series, N° 814).

36 Requirements for Hepatitis B Vaccines made by Recombinant DNA Techniques, in WHO Expert Committee on Biological Standardization. Thirty-ninth Report. Geneva, World Health Organization, 1989, Annex 2 (WHO Technical Report Series, N° 786).

37 Peebles PT: An in vitro focus-induction assay for xenotropic murine leukemia virus, feline leukemia virus C, and the feline-primates viruses RD-114/CC/M-7, Virology 1975;67:288-291.

38 Sammerfelt MA, Weiss RA: Receptor interference groups of 20 retroviruses plating on human cells. Virology 1990;176:58-69.

39 Lander MR, Chattopadhyay SK: A *Mus dunni* cell line that lacks sequences closely related to endogenous murine leukemia viruses and can be infected by ecotropic, amphotropic, xenotropic, and mink cell focus-forming viruses. Journal of Virology 1984;52:695-698.

40 Hartley RH, Rowe WP: Clonal cell lines from a feral mouse embryo which lack host-range restrictions for murine leukemia viruses. Virology 1975;65:128-134.

41 Bassin RH et al: Normal DBA/2 mouse cells synthesize a glycoprotein which interferes with MCF virus infection. Virology 1982;123:139-151.

42 Bassin RH, Tuttle N, Fischinger PJ: Rapid cell culture assay technic for murine leukemia viruses. Nature 1971;229:564-566.

43 Analysis of the expression construct in cells used for production of a DNA- derived protein product. Geneva, International Conference on Harmonisation, 1995 (unpublished document available on request from International Conference on Harmonisation Secretariat, International Federation of Pharmaceutical Manufacturers Associations, 30 rue de St-Jean, 1211 Geneva 18, Switzerland).

APPENDIX

Validation of viral elimination from monoclonal antibodies and biologicals prepared using recombinant DNA technology (except viral vaccines)

Traditionally, cell lines used as cell substrates for the production of biologicals have been tested to ensure the absence of contamination with adventitious viruses. As continuous cell lines have been introduced, it has become necessary to approve for production cell lines that produce virus-like particles and even infectious viruses. These efforts have resulted in an enhanced understanding of the significance of virus-like particles in cell lines and have demonstrated that certain findings, such as the presence of intracisternal A particles, are only of remote theoretical concern. As experience has been gained with monoclonal antibodies produced in cell lines that produce murine retroviruses, evidence has accumulated that such products can be safe, and methods have been developed to minimize both the potential for contamination of the products with retroviruses and the theoretical risk associated with such contamination. In particular, manufacturers have used manufacturing procedures that include steps that inactivate and/or remove viruses from the product, and have performed studies to validate the effectiveness of these procedures. When the manufacturing process is known to eliminate significantly more virus than is present in the unprocessed bulk, and when the purified product is tested for the presence of virus, there is reasonable assurance of freedom from contamination.

Validation studies assist in the quantification of risk, but do not of themselves prove absence of risk. They are relevant for the evaluation of production using cell lines potentially carrying any type of virus (e.g. Epstein-Barr virus, papillomavirus), but risk assessment also includes consideration of the type of virus and the potential use of the product. Validation studies are not a means of demonstrating that introduction of an adventitious virus during manufacture is acceptable. Validation is accomplished by evaluating the ability of down-stream processing steps to remove and/or inactivate virus from the bulk harvest: virus is added to test the efficacy of selected steps in a scaled-down model of the manufacturing process.

Design

The design of procedures to validate the elimination of virus during processing should take into account the following variables.

Selection of appropriate virus or viruses

The virus or viruses to be used may be the virus which is known or suspected to contaminate the cell line, or it may be a model virus (or viruses) selected because of its similarity to the virus of concern and because of practical considerations, such as availability of material of high titre and the ease of assay. The viral contaminant may be added in a labelled (i.e. radioactive) or non-labelled form. It may be necessary to use more than one virus when, for example, the use of a single virus does not provide an adequate basis for the evaluation of the purification process.

Scaled-down manufacturing system

If a scaled-down model of the purification process is used for validation, it should accurately reflect the actual manufacturing process. Bed height, flow rate, flow-rate to bed-height ratio, types of buffer, pH, and the concentration of protein, buffer and product should all be evaluated, and equivalence to the full-scale manufacturing system demonstrated.

Analysis of step-wise elimination of virus

In many cases it is desirable to evaluate the individual contribution to virus elimination of different manufacturing steps. Sufficient virus should be present before each critical step so that an adequate evaluation of the effectiveness of each step is obtained. In some cases, the addition of high-titre virus to the unpurified bulk and the testing of its concentration between steps will be sufficient. In other cases, the addition of virus to material during the manufacturing process will also be necessary. The virus titre should be determined before and after each tested step.

Determining physical removal or inactivation

The type of contribution (removal or inactivation) of each step should be identified by determining, when feasible, what portion of the reduction in titre is due to virus inactivation and what portion is due to physical removal of the virus from the product.

Kinetics of inactivation

In some cases, the kinetics of virus inactivation at the critical inactivation step should be determined. This is particulary important where the virus is known to be a human pathogen and it is necessary to design a completely effective inactivation process.

Estimation of combined effects

The combined effect of each tested step on the reduction of virus titre should be calculated in order to establish the total virus inactivation/ removal capacity of the purification procedure. Where a process involves several steps that achieve a reduction in titre by the same mechanism, unless otherwise justified the results of only one such step should be considered in calculating the overall titre reduction.

Regeneration of columns

When chromatographic procedures are used for virus elimination, it is critical that validation studies should employ columns that are representative of those actually used in manufacturing. Routine procedures for the regeneration of columns should be such that the design of the validation study is relevant to the manufacturing process.

Specific precautions

- Validation usually takes place outside the manufacturing facility in order to prevent possible viral contamination of the facility.
- Care should be taken in preparing virus preparations to avoid the aggregation of viral particles. This may facilitate physical removal and hinder inactivation, thereby reducing comparability with the actual manufacturing process.
- The virus preparation to be added to the product should constitute a small volume so as not to dilute or change the characteristics of the product.
- Care should be taken to avoid even small differences in, for example, buffers, media or reagents as these can substantially affect virus clearance and comparability with the manufacturing process.
- As virus removal/inactivation is time dependent, the amount of time the product remains in a buffer solution or on a chromatography column should reflect the conditions of the full-scale manufacturing process.
- Buffers and product should be evaluated independently for interference with the assays used to determine the virus titre, since these components may adversely affect the indicator cells. If the buffer solutions are toxic for the indicator cells, dilution, adjustment of pH or dialysis of the virus-containing buffer may be necessary. If the biological product itself has an antiviral activity, it may be possible to perform the validation study without the product, although omission of product or substitution of a similar protein without antiviral activity could affect behaviour of the virus in some manufacturing steps.
- Many purification schemes repeatedly use the same or similar buffers or chromatography columns. The effects of this approach should be taken into account when analysing the data. The effectiveness of virus removal/inactivation by a particular process may vary according to the stage in manufacture at which it is used.

Interpretation

The purpose of a validation study is to show that a process, when conducted according to standard operating procedures, will reliably produce a certain result. For viral contaminants, it is important to show that not only is the virus removed and/or inactivated, but also that there is excess capacity for this built into the purification process that will assure an appropriate level of safety for the final product. It is recommended that a purification process should include at least one viral inactivation step when infectious virus is known to be routinely present in the unpurified bulk product.

The following potential limitations of validation studies for virus removal or inactivation should be addressed when interpreting results.

- A model virus may not behave identically to relevant potential viral contaminants.
- The full-scale manufacturing process may differ from the scaled-down process used for validation purposes.

- Unrecognized differences in the materials or procedures used for validation as compared with those used for manufacturing may overestimate virus removal or inactivation.
- The effects of repeated steps, particularly of those with little individual effect, may not be additive, and summation of the effects of such steps may result in overestimation.
- The efficacy of chromatography colums and other devices used in the purification scheme may change on repeated use.

Statistics

Validation studies should analyse data statistically. Validation studies should be duplicated, and the statistical variation within and between them evaluated.

The design of the validation study should be statistically valid, i.e. it should be capable of supporting the conclusions reached.

Brown F, Griffiths E, Horaud F, Petricciani JC (eds): Safety of Biological Products Prepared from Mammalian Cell Culture.
Dev Biol Stand. Basel, Karger, 1998, vol 93, p 173

International Conference on Harmonisation (ICH)

Regulatory Documents

ICH Topic Q5A
Quality of Biotechnological Products: Viral Safety Evaluation of Biotechnology Products Derived from Cell Lines of Human or Animal Origin

Note for Guidance on Quality of Biotechnological Products: Viral Safety Evaluation of Biotechnology Products Derived from Cell Lines of Human or Animal Origin (CPMP/ICH/295/95)

Transmission to CPMP	**December 1995**
Transmission to interested parties	**December 1995**
Comments requested before	**June 1996**
Final approval by CPMP	**April 1997**
Date for coming into operation (for studies coming after)	**October 1997**

Brown F, Griffiths E, Horaud F, Petricciani JC (eds): Safety of Biological Products Prepared from Mammalian Cell Culture.
Dev Biol Stand. Basel, Karger, 1998, vol 93, pp 177-201

Quality of Biotechnological Products: Viral Safety Evaluation of Biotechnology Products Derived from Cell Lines of Human or Animal Origin

ICH Harmonised Tripartite Guideline

INTRODUCTION

This document is concerned with testing and evaluation of the viral safety of biotechnology products derived from characterised cell lines of human or animal origin (i.e. mammalian, avian, insect) and outlines data that should be submitted in the marketing application/registration package. For the purposes of this document the term virus excludes nonconventional transmissible agents like those associated with Bovine Spongiform Encephalopathy (BSE) and scrapie. Applicants are encouraged to discuss issues associated with BSE with the regulatory authorities.

The scope of the document covers products derived from cell cultures initiated from characterised cell banks. It covers products derived from in vitro cell culture, such as interferons, monoclonal antibodies and recombinant DNA-derived products including recombinant subunit vaccines, and also includes products derived from hybridoma cells grown in vivo as ascites. In this latter case, special considerations apply and additional information on testing cells propagated in vivo is contained in Appendix 1. Inactivated vaccines, all live vaccines containing self-replicating agents, and genetically engineered live vectors are excluded from the scope of this document.

The risk of viral contamination is a feature common to all biotechnology products derived from cell lines. Such contamination could have serious clinical consequences and can arise from the contamination of the source cell lines themselves (cell substrates) or from adventitious introduction of virus during production. To date, however, biotechnology products derived from cell lines have not been implicated in the transmission of viruses. Nevertheless, it is expected that the safety of these products with regard to viral contamination can be reasonably assured only by the application of a virus testing program and assessment of virus removal and inactivation achieved by the manufacturing process, as outlined below.

Three principal, complementary approaches have evolved to control the potential viral contamination of biotechnology products:

Appendix 1: Products derived from characterised cell banks which were subsequently grown in vivo.

For products manufactured from fluids harvested from animals inoculated with cells from characterised banks, additional information regarding the animals should be provided. Whenever possible, animals used in the manufacture of biotechnological/biological products should be obtained from well defined, specific pathogen-free colonies. Adequate testing for appropriate viruses, such as those listed in Table 3, should be performed. Quarantine procedures for newly arrived as well as diseased animals should be described, and assurance provided that all containment, cleaning and decontamination methodologies employed within the facility are adequate to contain the spread of adventitious agents. This may be accomplished through the use of a sentinel program. A listing of agents for which testing is performed should also be included. Veterinary support services should be available on-site or within easy access. The degree to which the vivarium is segregated from other areas of the manufacturing facility should be described. Personnel practices should be adequate to ensure safety.

Procedures for the maintenance of the animals should be fully described. These would include diet, cleaning and feeding schedules, provisions for periodic veterinary care if applicable, and details of special handling that the animals may require once inoculated. A description of the priming regimen(s) for the animals, the preparation of the inoculation and the site and route of inoculation should also be included.

The primary harvest material from animals may be considered an equivalent stage of manufacture to unprocessed bulk harvest from a bioreactor. Therefore, all testing considerations previously outlined in section 4. of this document should apply. In addition, the manufacturer should assess the bioburden of the unprocessed bulk, determine whether the material is free of mycoplasma, and perform species-specific assay(s) as well as in vivo testing in adult and suckling mice.

a) selecting and testing cell lines and other raw materials, including media components, for the absence of undesirable viruses which may be infectious and/or pathogenic for humans;
b) assessing the capacity of the production processes to clear infectious viruses;
c) testing the product at appropriate steps of production for absence of contaminating infectious viruses.

All testing suffers from the inherent limitation of quantitative virus assays, i.e., that the ability to detect low viral concentrations depends for statistical reasons on the size of the sample. Therefore, no single approach will necessarily establish the safety of a product. Confidence that infectious virus is absent from the final product will in many instances not be derived solely from direct testing for their presence, but also from a demonstration that the purification regimen is capable of removing and/or inactivating the viruses.

The type and extent of viral tests and viral clearance studies required at different steps of production will depend on various factors and should be considered on a case-by-case and step-by-step basis. The factors that should be taken into account include the extent of cell bank characterisation and qualification, the nature of any viruses detected, culture medium constituents, culture methods, facility and equipment design, the results of viral tests after cell culture, the ability of the process to clear viruses, and the type of product and its intended clinical use.

The purpose of this document is to provide a general framework for virus testing, experiments for the assessment of viral clearance and a recommended approach for the design of viral tests and viral clearance studies. Related information is described in the appendices and selected definitions are provided in the glossary.

The manufacturers should adjust the recommendations presented here to their specific product and its production process. The approach used by manufacturers in their overall strategy for ensuring viral safety should be explained and justified. In addition to the detailed data which is provided, an overall summary of the viral safety assessment would be useful in facilitating the review by regulatory authorities. This summary should contain a brief description of all aspects of the viral safety studies and strategies used to prevent virus contamination as they pertain to this document.

Potential Sources of Virus Contamination

Viral contamination of biotechnology products may arise from the original source of the cell lines or from adventitious introduction of virus during production processes.

Viruses that could occur in the Master Cell Bank (MCB)

Cells may have latent or persistent virus infection (e.g., herpesvirus) or endogenous retrovirus which may be transmitted vertically from one cell generation to the next, since the viral genome persists within the cell. Such viruses may be constitutively expressed or may unexpectedly become expressed as an infectious virus.

Viruses can be introduced into the MCB by several routes such as: 1) derivation of cell lines from infected animals; 2) use of virus to establish the cell line; 3) use of contaminated biological reagents such as animal serum components; 4) contamination during cell handling.

Adventitious viruses that could be introduced during production

Adventitious viruses can be introduced into the final product by several routes including, but not limited to, the following: 1) the use of contaminated biological reagents such as animal serum components; 2) the use of a virus for the induction of expression of specific genes encoding a desired protein; 3) the use of a contaminated reagent, such as a monoclonal antibody affinity column; 4) the use of a contaminated excipient during formulation; 5) contamination during cell and medium handling. Monitoring of cell culture parameters can be helpful in the early detection of potential adventitious viral contamination.

Cell Line Qualification: Testing for Viruses

An important part of qualifying a cell line for use in the production of a biotechnology product is the appropriate testing for the presence of virus.

Suggested virus tests for MCB, Working Cell Bank (WCB) and cells at the limit of in vitro cell age used for production

Table 1 shows an example of virus tests to be performed once only at various cell levels, including MCB, WCB and cells at the limit of in vitro cell age used for production.

Master Cell Bank

Extensive screening for both endogenous and non-endogenous viral contamination should be performed on the MCB. For heterohybrid cell lines in which one or more partners are human or non-human primate in origin, tests should be performed in order to detect viruses of human or non-human primate origin as viral contamination arising from these cells may pose a particular hazard.

Testing for non-endogenous viruses should include in vitro and in vivo inoculation tests and any other specific tests, including species-specific tests such as the

Table 1: Virus tests to be performed once at various cell levels.

	MCB	WCB[a]	Cell at the limit
Tests for retroviruses and other endogenous viruses			
Infectivity	+	–	+
Electron microscopy[c]	+[c]	–	+[c]
Reverse transcriptase[d]	+[d]	–	+[d]
Other virus-specific tests[e]	as appropriate [e]	–	as appropriate [e]
Tests for non-endogenous or adventitious virus test			
In vitro assays	+	–[f]	+
In vivo assays	+	–[f]	+
Antibody production tests[g]	+[g]	–	–
Other virus-specific tests[h]	+[h]	–	–

[a] See text – Section 3.1.2

[b] Cells at the limit; Cells at the limit of in vitro cell age used for production (See text – Section 3.1.3).

[c] May also detect other agents.

[d] Not necessary if positive by retrovirus infectivity test.

[e] As appropriate for cell lines which are known to have been infected by such agents.

[f] For the first WCB, this test should be performed on cells at the limit of in vitro cell age, generated from that WCB; for WCBs subsequent to the first WCB, a single in vitro and in vivo test can be done either directly on the WCB or on cells at the limit of in vitro cell age.

[g] e.g., MAP, RAP, HAP – Usually applicable for rodent cell lines.

[h] e.g., Tests for cell lines derived from human, non-human primate or other cell lines as appropriate.

mouse antibody production (MAP) test, that are appropriate, based on the passage history of the cell line, to detect possible contaminating viruses.

Working Cell Bank

Each WCB as a starting cell substrate for drug production should be tested for adventitious virus either by direct testing or by analysis of cells at the limit of in vitro cell age, initiated from the WCB. When appropriate non-endogenous virus tests have been performed on the MCB and cells cultured up to or beyond the limit of in vitro cell age have been derived from the WCB and used for testing for the presence of adventitious viruses, similar tests need not be performed on the initial WCB. Antibody production tests are usually not necessary for the WCB. An alternative approach in which full tests are carried out on the WCB rather than on the MCB would also be acceptable.

Cells at the limit of in vitro cell age used for production

The limit of in vitro cell age used for production should be based on data derived from production cells expanded under pilot-plant scale or commercial-scale conditions to the proposed in vitro cell age or beyond. Generally, the production cells are obtained by expansion of the WCB; the MCB could also be used to prepare the production cell. Cells at the limit of in vitro cell age should be evaluated once for those endogenous viruses that may have been undetected in the MCB and WCB. The performance of suitable tests (e.g., in vitro and in vivo at least once on cells at the limit of in vitro cell age used for production would provide further assurance that the production process is not prone to contamination by adventitious virus. If any adventitious viruses are detected at this level, the process should be carefully checked in order to determine the cause of the contamination, and completely redesigned if necessary.

Recommended viral detection and identification assays

Numerous assays can be used for the detection of endogenous and adventitious viruses. Table 2 outlines examples for these assays. They should be regarded as assay protocols recommended for the present, but the list is not all-inclusive or definitive. Since the most appropriate techniques may change with scientific progress, proposals for alternative techniques, when accompanied by adequate supporting data, may be acceptable. Manufacturers are encouraged to discuss these alternatives with the regulatory authorities. Other tests may be necessary depending on the individual case. Assays should include appropriate controls to ensure adequate sensitivity and specificity. Wherever a relatively high possibility of the presence of a specific virus can be predicted from the species of origin of the cell substrate, specific tests and/or approaches may be necessary. If the cell line used for production is of human or non-human primate origin, additional tests for human viruses, such as those causing immunodeficiency diseases and hepatitis, should be performed unless otherwise justified. The polymerase chain reaction (PCR) may be appropriate for detection of

Table 2: Examples of the use and limitations of assays which may be used to test for virus.

Test	Test article	Detection capability	Detection limitation
Antibody production	Lysate of cells and their culture medium	Specific viral antigens	Antigens not infectious for animal test system
In vivo virus screen	Lysate of cells and their culture medium	Broad range of viruses pathogenic for humans	Agents failing to replicate or produce diseases in the test system
In vitro virus screen for:		Broad range of viruses pathogenic for humans	Agents failing to replicate or produce diseases in the test system
1. Cell bank characterisation 2. Production screen	1. Lysate of cells and their culture medium (for co-cultivation, intact cells should be in the test article) 2. Unprocessed bulk harvest or lysate of cells and their cell culture medium from the production reactor		
TEM on :		Virus and virus-like particles	Qualitative assay with assessment of identity
1. Cell substrate 2. Cell culture supernatant	1. Viable cells 2. Cell-free culture supernatant		
Reverse transcriptase (RT)	Cell-free culture supernatant	Retroviruses and expressed retroviral RT	Only detects enzymes with optimal activity under preferred conditions. Interpretation may be difficult due to presence of cellular enzymes; background with some concentrated samples
Retrovirus (RV) infectivity	Cell-free culture supernatant	Infectious retroviruses	RV failing to replicate or form discrete foci or plaques in the chosen test system
Cocultivation 1. Infectivity endpoint 2. TEM endpoint 3. RT endpoint	Viable cells	Infectious retroviruses	RV failling to replicate 1. See above under RV infectivity 2. See above under TEM 3. See above under RT
PCR (Polymerase chain reaction)	Cells, culture fluid and other materials	Specific virus sequences	Primer sequences must be present. Does not indicate whether virus is infectious

a In addition, difficult to distinguish test article from indicator cells.

sequences of these human viruses as well as for other specific viruses. The following is a brief description of a general framework and philosophical background within which the manufacturer should justify what was done.

Tests for retroviruses

For the MCB and for cells cultured up to or beyond the limit of in vitro cell age used for production, tests for retroviruses, including infectivity assays in sensitive cell cultures and electron microscopy (EM) studies, should be carried out. If infectivity is not detected and no retrovirus or retrovirus-like particles have been observed by EM, reverse transcriptase (RT) or other appropriate assays should be performed to detect retroviruses which may be non-infectious. Induction studies have not been found to be useful.

In vitro assays

In vitro tests are carried out by the inoculation of a test article (see Table 2) into various susceptible indicator cell cultures capable of detecting a wide range of human and relevant animal viruses. The choice of cells used in the test is governed by the species of origin of the cell bank to be tested, but should include a human and/or a non-human primate cell line susceptible to human viruses. The nature of the assay and the sample to be tested are governed by the type of virus which may possibly be present based on the origin or handling of the cell. Both cytopathic and hemadsorbing viruses should be sought.

In vivo assays

A test article (see Table 2) should be inoculated into animals, including suckling and adult mice, and in embryonated eggs to reveal viruses that cannot grow in cell cultures. Additional animal species may be used depending on the nature and source of the cell lines being tested. The health of the animals should be monitored and any abnormality should be investigated to establish the cause of the illness.

Antibody production tests

Species-specific viruses present in rodent cell lines may be detected by inoculating test article (see Table 2) into virus-free animals, and examining the serum antibody level or enzyme activity after a specified period. Examples of such tests are the mouse antibody production (MAP) test, rat antibody production (RAP) test, and hamster antibody production (HAP) test. The viruses currently screened for in the antibody production assays are discussed in Table 3.

Acceptability of cell lines

It is recognised that some cell lines used for the manufacture of product will contain endogenous retroviruses, other viruses or viral sequences. In such circum-

Table 3: Virus detected in antibody production tests.

MAP	HAP	RAP
Ectromelia Virus[2,3]	Lymphocytic Choriomeningitis Virus (LCM)[1,3]	Hantaan Virus[1,3]
Hantaan Virus[1,3]	Pneumonia Virus of Mice (PVM)[2,3]	Kilham Rat Virus (KRV)[2,3]
K Virus[2]	Reovirus Type 3 (Reo3)[1,3]	Mouse Encephalomyelitis Virus (Theilers, GDVII)[2]
Lactic Dehydrogenase Virus (LDM)[1,3]	Sendai Virus[1,3]	Pneumonia Virus of Mice (PVM)[2,3]
Lymphocytic Choriomeningitis Virus (LCM)[1,3]	SV5	Rat Coronavirus [2]
Minute Virus of Mice[2,3]		Reovirus Type 3 (Reo3)[1,3]
Mouse Adenovirus (MAV)[2,3]		Sendai Virus[1,3]
Mouse Cytomegalovirus (MCMV)[2,3]		Sialoacryoadenitis Virus (SDAV)[2]
Mouse Encephalomyelitis Virus (Theilers, GDVII)[2]		Toolan Virus (HI)[2,3]
Mouse Hepatitis Virus (MHV)[2]		
Mouse Rotavirus (EDIM)[2,3]		
Pneumonia Virus of Mice (PVM)[2,3]		
Polyoma Virus[2]		
Reovirus Type 3 (Reo3)[1,3]		
Sendai Virus[1,3]		
Thymic Virus[2]		

1 Viruses for which there is evidence of capacity for infecting humans or primates.
2 Viruses for which there is no evidence of capacity for infecting humans.
3 Virus capable of replicating in vitro in cells of human or primate origin.

stances, the action plan recommended for manufacture is described in Section 5 of this document. The acceptability of cell lines containing viruses other than endogenous retroviruses will be considered on an individual basis by the regulatory authorities, by taking into account a risk/benefit analysis based on the benefit of the product and its intended clinical use, the nature of the contaminating viruses, their potential for infecting humans or for causing disease in humans, the purification process for the product (e.g., viral clearance evaluation data), and the extent of the virus tests conducted on the purified bulk.

Testing for Viruses in Unprocessed Bulk

The unprocessed bulk constitutes one or multiple pooled harvests of cells and culture media. When cells are not readily accessible (e.g., hollow fiber or similar systems), the unprocessed bulk would constitute fluide harvested from the fermenter. A representative sample of the unprocessed bulk, removed from the production reactor prior to further processing, represents one of the most suitable levels at which the possibility of adventitious virus contamination can be determined with a high probability of detection. Appropriate testing for viruses should be performed at the unprocessed bulk level unless virus testing is made more sensitive by initial partial processing (e.g., unprocessed bulk may be toxic in test cell cultures, whereas partially processed bulk may not be toxic).

In certain instances it may be more appropriate to test a mixture consisting of both intact and disrupted cells and their cell culture supernatants removed from the production reactor prior to further processing. Data from at least 3 lots of unprocessed bulk at pilot-plant scale or commercial scale should be submitted as part of the marketing application/registration package.

It is recommended that manufacturers develop programs for the ongoing assessment of adventitious viruses in production batches. The scope, extent and frequency of virus testing on the unprocessed bulk should be determined by taking several points into consideration including the nature of the cell lines used to produce the desired products, the results and extent of virus tests performed during the qualification of the cell lines, the cultivation method, raw material sources and results of viral clearance studies. In vitro screening tests, using one or several cell lines, are generally employed to test unprocessed bulk. If appropriate, a PCR test or other suitable methods may be used.

Generally, harvest material in which adventitious virus has been detected should not be used to manufacture the product. If any adventitious viruses are detected at this level, the process should be carefully checked to determine the cause of the contamination, and appropriate actions taken.

Rationale and Action Plan for Viral Clearance Studies and Virus Tests on Purified Bulk

It is important to design the most relevant and rational protocol for virus tests from the MCB level, through the various steps of drug production, to the final product including evaluation and characterisation of viral clearance from unprocessed bulk. The evaluation and characterisation of viral clearance plays a critical role in this scheme. The goal should be to obtain the best reasonable assurance that the product is free of virus contamination.

In selecting viruses to use for a clearance study, it is useful to distinguish between the need to evaluate processes for their ability to clear viruses that are known to be present and the desire to estimate the robustness of the process by characterising the clearance of non-specific «model» viruses (described later). Definitions of «relevant», specific and non-specific «model» viruses are given in the glossary. Process evaluation requires knowledge of how much virus may be present in the process, such as the unprocessed bulk, and how much can be cleared in order to assess product safety. Knowledge of the time dependence for inactivation procedures is helpful in assuring the effectiveness of the inactivation process. When evaluating clearance of known contaminants, in-depth time-dependent inactivation studies, demonstration of reproducibility of inactivation/removal, and evaluation of process parameters will be needed. When a manufacturing process is characterised for robustness of clearance using non-specific «model» viruses, particular attention should be paid to non-enveloped viruses in the study design. The extent of viral clearance characterisation studies may be influenced by the results of tests on cell lines and unprocessed bulk. These studies should be performed as described below (section 6.).

Table 4 presents an example of an action plan, in terms of process evaluation and characterisation of viral clearance as well as virus tests on purified bulk, in response to the results of virus tests on cells and/or the unprocessed bulk. Various cases are considered. In all cases, characterisation of clearance using nonspecific «model» viruses should be performed. The most common situations are Cases A and B. Production systems contaminated with a virus other than a rodent retrovirus are normally not used. Where there are convincing and well justified reasons for drug production using a cell line from Cases C, D or E, these should be discussed with the regulatory authorities. With Cases C, D and E it is important to have validated effective steps to inactivate/remove the virus in question from the manufacturing process.

Case A: Where no virus, virus-like particle or retrovirus-like particle has been demonstrated in the cells or the unprocessed bulk, virus removal and inactivation studies should be performed with non-specific «model» viruses as previously stated.

Case B: Where only a rodent retrovirus (or a retrovirus-like particle which is believed to be non-pathogenic, such as rodent A- and R-type particles) is present, process evaluation using a specific «model» virus, such as a murine leukemia virus, should be performed. Purified bulk should be tested using suitable methods having high specificity and sensitivity for the detection of the virus in question. For marketing authorisation, data from at least 3 lots of purified bulk at pilot-plant scale or commercial scale should be provided. Cell lines such as CHO, C127, BHK and murine hybridoma cell lines have frequently been used as substrates for drug production with no reported safety problems related to viral contamination of the products. For these cell lines in which the endogenous particles have been extensively characterised and clearance has been demonstrated, it is not usually necessary to assay for the presence of the non-infectious particles in purified bulk. Studies with non-specific «model» viruses, as in Case A, are appropriate.

Case C: When the cells or unprocessed bulk are known to contain a virus, other than a rodent retrovirus, for which there is no evidence of capacity for infecting humans, (such as those identified by footnote 2 in Table 3, except rodent retroviruses (Case B)), virus removal and inactivation evaluation studies should use the identified virus. If it is not possible to use the identified virus, «relevant» or specific «model» viruses should be used to demon-

Table 4: Action plan for process evaluation of viral clearance and virus tests on products.

	Case A	Case B	Case C[2]	Case D[2]	Case E[2]
Status					
Presence of virus[1]	–	–	+	+	(+)[3]
Virus-like particles[1]	–	–	–	–	(+)[3]
Retrovirus like particles[1]	–	+	–	–	(+)[3]
Virus identified	not applicable	+	+	+	–
Virus pathogenic for humans	not applicable	–[4]	–[4]	+	unknown
Action					
Process characterisation of viral clearance using non-specific «model» viruses	yes[5]	yes[5]	yes[5]	yes[5]	yes[7]
Process evaluation of viral clearance using «relevant» or specific «model» viruses	no	yes[6]	yes[6]	yes[6]	yes[7]
Test for virus in purified bulk	not applicable	yes[8]	yes[8]	yes[8]	yes[8]

1 Results of virus tests for the cell substrate and/or at the unprocessed bulk level. Cell cultures used for production which are contaminated with viruses will generally not be acceptable. Endogenous viruses (such as retroviruses) or viruses that are an integral part of the MCB may be acceptable if appropriate viral clearance evaluation procedures are followed.

2 The use of source material which is contaminated with viruses, whether or not they are known to be infectious and/or pathogenic in humans, will only be permitted under very exceptional circumstances.

3 Virus has been observed by either direct or indirect methods.

4 Believed to be non-pathogenic.

5 Characterisation of clearance using non-specific «model» viruses should be performed.

6 Process evaluation for «relevant» viruses or specific «model» viruses should be performed.

7 See text under Case E.

8 The absence of detectible virus should be confirmed for purified bulk by means of suitable methods having high specificity and sensitivity for the detection of the virus in question. For the purpose of marketing authorisation, data from at least 3 lots of purified bulk manufactured at pilot-plant or commercial scale should be provided. However for cell lines such as CHO cells for which the endogenous particles have been extensively characterised and adequate clearance has been demonstrated, it is not usually necessary to assay for the presence of the non-infectious particles in purified bulk.

strate acceptable clearance. Time-dependent inactivation for identified (or «relevant» or specific «model») viruses at the critical inactivation step(s) should be obtained as part of process evaluation for these viruses. Purified bulk should be tested using suitable methods having high specificity and sensitivity for the detection of the virus in question. For the purpose of marketing authorisation, data from at least 3 lots of purified bulk manufactured at pilot-plant scale or commercial scale should be provided.

Case D: Where a known human pathogen, such as those indicated by footnote 1 in Table 3, is identified, the product may be acceptable only under exceptional circumstances. In this instance, it is recommended that the identified virus be used for virus removal and inactivation evaluation studies and specific methods with high specificity and sensitivity for the detection of the virus in question be employed. If it is not possible to use the identified virus, «relevant» and/or specific «model» viruses (described later) should be used. The process should be shown to achieve the removal and inactivation of the selected viruses during the purification and inactivation processes. Time-dependent inactivation data for the critical inactivation step(s) should be obtained as part of process evaluation. Purified bulk should be tested using suitable methods having high specificity for the detection of the virus in question. For the purpose of marketing authorisation, data from at least 3 lots of purified bulk manufactured at pilot-plant scale or commercial scale should be provided.

Case E: When a virus, which cannot be classified by currently available methodologies, is detected in cells or unprocessed bulk, the product is usually considered unacceptable since the virus may prove to be pathogenic. In the very rare case where are convincing and well justified reasons for drug production using such a cell line, this should be discussed with the regulatory authorities before proceeding further.

Evaluation and Characterisation of Viral Clearance Procedures

Evaluation and characterisation of the virus removal and/or inactivation procedures plays an important role in establishing the safety of biotechnology products. Many instances of contamination in the past occurred with agents whose presence was not known or even suspected, and though this happened to biological products derived from various source materials other than fully characterised cell lines, assessment of viral clearance will provide a measure of confidence that any unknown, unsuspected and harmful viruses may be removed. Studies should be carried out in a manner that is well documented and controlled.

The objective of viral clearance studies is to assess process step(s) that can be considered to be effective in inactivating/removing viruses and to estimate quantitatively the overall level of virus reduction obtained by the process. This should be achieved by the deliberate addition («spiking») of significant amounts of a virus to the crude material and/or to different fractions obtained during the various process steps and demonstrating its removal or inactivation during the subsequent steps. It is not necessary to evaluate or characterise every step of a manufacturing process if adequate clearance is demonstrated by the use of fewer steps. It should be borne in mind that other steps in the process may have an indirect effect on the viral inactivation/removal achieved. Manufacturers should explain and justify the approach used in studies for evaluating virus clearance.

The reduction of virus infectivity may be achieved by removal of virus particles or by inactivation of viral infectivity. For each production step assessed, the possible mechanism of loss of viral infectivity should be described with regard to whether it is due to inactivation or removal. For inactivation steps, the study should be planned in such a way that samples are taken at different times and an inactivation curve constructed (see section 6.2.5).

Viral clearance evaluation studies are performed to demonstrate the clearance of a virus known to be present in the MCB and/or to provide some level of assurance that adventitious viruses which could not be detected, or might gain access to the production process, would be cleared. Reduction factors are normally expressed on a logarithmic scale which implies that, while residual virus infectivity will never be reduced to zero, it may be greatly reduced mathematically.

In addition to clearance studies for viruses known to be present, studies to characterise the ability to remove and/or inactivate other viruses should be conducted. The purpose of studies with viruses, exhibiting a range of biochemical and biophysical properties, that are not known or expected to be present, is to characterise the robustness of the procedure rather than to achieve a specific inactivation or removal goal. A demonstration of the capacity of the production process to inactivate or remove viruses is desirable (see Section 6.3.). Such studies are not performed to evaluate a specific safety risk. Therefore, a specific clearance value need not be achieved.

The choice of viruses for the evaluation and characterisation of viral clearance

Viruses for clearance evaluation and process characterisation studies should be chosen to resemble viruses which may contaminate the product and to represent a wide range of physico-chemical properties in order to test the ability of the system to eliminate viruses in general. The manufacturer should justify the choice of viruses in accordance with the aims of the evaluation and characterisation study and the guidance provided in this guideline.

«Relevant» viruses and «model» viruses

A major issue in performing a viral clearance study is to determine which viruses should be used. Such viruses fall into three categories «relevant» viruses, specific «model» viruses and non-specific «model» viruses.

«Relevant» viruses are viruses used in process evaluation of viral clearance studies which are either the identified viruses, or of the same species as the viruses that are known, or likely to contaminate the cell substrate or any other reagents or materials used in the production process. The purification and/or inactivation process should demonstrate the capability to remove and/or inactivate such viruses. When a «relevant» virus is not available or when it is not well adapted to process evaluation of viral clearance studies (e.g. it cannot be grown in vitro to sufficiently high titres), a specific «model» virus should be used as a substitute. An appropriate specific «model» virus may be a virus which is closely related to the known or suspected virus (same genus or family), having similar physical and chemical properties to the observed or suspected virus.

Cell lines derived from rodents usually contain endogenous retrovirus particles or retrovirus-like particles, which may be infectious (C-type particles) or non-infectious (cytoplasmic A- and R-type particles). The capacity of the manufacturing process to remove and/or inactivate rodent retroviruses from products obtained from such cells should be determined. This may be accomplished by using a murine leukemia virus, a specific «model» virus in the case of cells of murine origin. When human cell lines secreting monoclonal antibodies have been obtained by the immortalisation of B lymphocytes by Epstein-Barr Virus (EBV), the ability of the manufacturing process to remove and/or inactivate a herpes virus should be determined. Pseudorabies virus may also be used as a specific «model» virus.

When the purpose is to characterise the capacity of the manufacturing process to remove and/or inactivate viruses in general, i.e. to characterise the robustness of the clearance process, viral clearance characterisation studies should be performed with non-specific «model» viruses with differing properties. Data obtained from studies with «relevant» and/or specific «model» viruses may also contribute to this assessment. It is not necessary to test all types of viruses. Preference should be given to viruses that display a significant resistance to physical and/or chemical treatments. The results obtained for such viruses provide useful information about the ability of the production process to remove and/or inactivate viruses in general. The choice and number of viruses used will be influenced by the quality and characterisation of the cell lines and the production process.

Examples of useful «model» viruses representing a range of physico-chemical structures and examples of viruses which have been used in viral clearance studies are given in Appendix 2 and Table A- 1.

Other considerations

Additional points to be considered are as follows:

a) Viruses which can be grown to high titre are desirable, although this may not always be possible.

b) There should be an efficient and reliable assay for the detection of each virus used, for every stage of manufacturing that is tested.

c) Consideration should be given to the health hazard which certain viruses may pose to the personnel performing the clearance studies.

Design and implications of viral clearance evaluation and characterisation studies

Facility and staff

It is inappropriate to introduce any virus into a production facility because of GMP constraints. Therefore, viral clearance studies should be conducted in a separate laboratory equipped for virological work and performed by staff with virological expertise in conjunction with production personnel involved in designing and preparing a scaled-down version of the purification process.

Appendix 2: The choice of viruses for viral clearance studies.

1.	**Examples of useful «model» viruses:**
	a. Non-specific "model" viruses representing a range of physico-chemical structures:
	– SV40 (Polyomavirus maccacae 1), human polio virus 1 (Sabin), animal parvovirus or some other small, non-enveloped viruses;
	– a parainfluenza virus or influenza virus, Sindbis virus or some other medium-to-large, enveloped, RNA viruses;
	– a herpes virus (e.g. HSV-1 or a pseudorabies virus), or some other medium-to-large, DNA viruses.
	These viruses are exemples only and their use is not mandatory.
	b. For rodent cell substrates murine retroviruses are commonly used as specific «model» viruses.
2.	**Examples of viruses which have been used in viral clearance studies**
	Several viruses which have been used in viral clearance studies are listed in Table A-1. However, since these are merely examples, the use of any of the viruses in the table is not mandatory and manufacturers are invited to consider other viruses, especially those which may be more appropriate for their individual production processes. Generally, the process should be assessed for its ability to clear at least three different viruses with differing characteristics.

Scaled-down production system

The validity of the scaling down should be demonstrated. The lever of purification of the scaled-down version should represent as closely as possible the production procedure. For chromatographic equipment, column bed-height, linear flow-rate, flow-rate-to-bed-volume ratio (i.e., contact time), buffer and gel types, pH, temperature, and concentration of protein, salt, and product should all be shown to be representative of commercial-scale manufacturing. A similar elusion profile should result. For other procedures, similar considerations apply. Deviations which cannot be avoided should be discussed with regard to their influence on the results.

Analysis of step-wise elimination of virus

When viral clearance studies are being performed, it is desirable to assess the contribution of more than one production step to virus elimination. Steps which are likely to clear virus should be individually assessed for their ability to remove and inactivate virus and careful consideration should be given to the exact definition of an individual step. Sufficient virus should be present in the material of each step to be tested so that an adequate assessment of the effectiveness of each step is obtained. Generally, virus should be added to in-process material of each step to be tested. In some cases, simply adding high titre virus to unpurified bulk and testing its concentration between steps will be sufficient. Where virus removal results from

Table A-1: Examples of viruses which have been used in virus clearance evaluation studies.

Virus	Family	Genus	Natural host	Genome	Env	Size (nm)	Shape	Resistance*
Vesicular stomatitis virus	Rhabdo	Vesiculovirus	Equine Bovine	RNA	Yes	70 x 150 nm	Bullet	Low
Parainfluenza virus	Paramyxo	Paramyxovirus	Various	RNA	Yes	100-200+	Pleo/Spher	Low
MuLV	Retro	Type C oncovirus	Mouse	RNA	Yes	80-110	Spherical	Low
Sindbis virus	Toga	Alphavirus	Human	RNA	Yes	60-70	Spherical	Low
BVDV	Flavi	Pestivirus	Bovine	RNA	Yes	50-70	Pleo-Spher	Low
Pseudorabies virus	Herpes		Swine	DNA	Yes	120-200	Spherical	Med
Poliovirus Sabin Type 1	Picorna	Enterovirus	Human	RNA	no	25-30	Icosahedral	Med
Encephalomyocarditis virus (EMC)	Picorna	Cardiovirus	Mouse	RNA	no	25-30	Icosahedral	Med
Reovirus 3	Reo	Orthoreovirus	Various	DNA	no	60-80	Spherical	Med
SV40	Papova	Polyomavirus	Monkey	DNA	no	40-50	Icosahedral	Very high
Parvoviruses (canine, porcine)	Parvo	Parvovirus	Canine Porcine	DNA	no	18-24	Icosahedral	Very high

* Resistance to physico-chemical treatments based on studies of production processes. Resistance is relative to the specific treatment and it is used in the context of the understanding of the biology of the virus and the nature of the manufacturing process. Actual results will vary according to the treatment.
These viruses are examples only and their use is not mandatory.

separation procedures, it is recommended that, if appropriate and if possible, the distribution of the virus load in the different fractions be investigated. When virucidal buffers are used in multiple steps within the manufacturing process, alternative strategies such as parallel spiking in less virucidal buffers, may be carried out as part of the overall process assessment. The virus titre before and after each step being tested should be determined. Quantitative infectivity assays should have adequate sensitivity and reproducibility and should be performed with sufficient replicates to ensure adequate statistical validity of the result. Quantitative assays not associated with infectivity may be used if justified. Appropriate virus controls should be included in all infectivity assays to ensure the sensitivity of the method. Also, the statistics of sampling virus when at low concentrations should be considered (Appendix 3).

Appendix 3: Statistical considerations for assessing virus assays.

Virus titrations suffer the problems of variation common to all biological assay systems. Assessment of the accuracy of the virus titrations and reduction factors derived from them and the validity of the assays should be performed to define the reliability of a study. The objective of statistical evaluation is to establish that the study has been carried out to an acceptable lever of virological competence.	
1.	Assay methods may be either quantal or quantitative. Quantal methods include infectivity assays in animals or in tissue-culture-infectious-dose (TCID) assays, in which the animal or cell culture is scored as either infected or not. Infectivity titres are then measured by the proportion of animals or culture infected. In quantitative methods, the infectivity measured varies continuously with the virus input. Quantitative methods include plaque assays where each plaque counted corresponds to a single infectious unit. Both quantal and quantitative assays are amenable to statistical evaluation.
2.	Variation can arise within an assay as a result of dilution errors, statistical effects and differences within the assay system which are either unknown or difficult to control. These effects are likely to be greater when different assay runs are compared (between assay variation) than when results within a single assay run are compared (within assay variation).
3.	The 95% confidence limits for results of within-assay variation normally should be on the order of $\pm$ 0.5 $\log_{10}$ of the mean. Within-assay variation can be assessed by standard textbook methods. Between-assay variation can be monitored by the inclusion of a reference preparation, the estimate of whose potency should be within approximately 0.5 $\log_{10}$ of the mean estimate established in the laboratory for the assay to be acceptable. Assays with longer precision may be acceptable with appropriate justification.
4.	The 95% confidence limits for the reduction factor observed should be calculated wherever possible in studies of clearance of «relevant» and specific «model» viruses. If the 95% confidence limits for the viral assays of the starting material are +s, and for the viral assays of the material after the step are +a, the 95% confidence limits for the reduction factor are $\pm \text{v}S^2 + a^2$
Probability of detection of viruses at low concentrations	
At low virus concentrations (e.g. in the range of 10 to 1000 infectious particles per litre) it is evident that a sample of a few millilitres may or may not contain infectious particles. The probability, *p*, that this sample does not contain infectious viruses is: $p = ((V-v)/V)^n$	

where V (litre) is the overall volume of the material to be tested, v (litre) is the volume of the sample and n is the absolute number of infectious particles statistically distributed in V.

If V >> v, this equation can be approximated by the Poisson distribution:

$$p = e^{-cv}$$

where c is the concentration of infectious particles per litre.

Or, c = ln *p*/–v

As an example, if a sample volume of 1 ml is tested, the probabilities *p* at virus concentrations ranging from 10 to 1000 infectious particles per litre are:

c	10	100	1000
p	0.99	0.90	0.37

This indicates that for a concentration of 1000 viruses per litre, in 37% of sampling, 1 ml will not contain a virus particle.
If only a portion of a sample is tested for virus and the test is negative, the amount of virus which would have to be present in the total sample in order to achieve a positive result should be calculated and this value taken into account when calculating a reduction factor. Confidence limits at 95% are desirable. However, in some instances, this may not be practical due to material limitations.

Determining physical removal versus inactivation

Reduction in virus infectivity may be achieved by the removal or inactivation of virus. For each production step assessed, the possible mechanism of loss of viral infectivity should be described with regard to whether it is due to inactivation or removal. If little clearance of infectivity is achieved by the production process, and the clearance of virus is considered to be a major factor in the safety of the product, specific or additional inactivation/removal steps should be introduced. It may be necessary to distinguish between removal and inactivation for a particular step, for example when there is a possibility that a buffer used in more than one clearance step may contribute to inactivation during each step; i.e., the contribution to inactivation by a buffer shared by several chromatographic steps and the removal achieved by each of these chromatographic steps should be distinguished.

Inactivation assessment

For assessment of viral inactivation, unprocessed crude material or intermediate material should be spiked with infectious virus and the reduction factor calculated. It should be recognised that virus inactivation is not a simple, first order reaction and is usually more complex, with a fast «phase 1» and a slow «phase 2». The study should, therefore, be planned in such a way that samples are taken at different times and an inactivation curve constructed. It is recommended that studies for inactivation include at least one time point less than the minimum exposure time and greater than zero, in addition to the minimum exposure time. Additional data are particularly important where the virus is a «relevant» virus known to be a human pathogen and an effective inactivation process is being designed. However, for inactivation studies in which non-specific «model» viruses are used or when specific «model» viruses are used as surrogates for virus particles such as the CHO intracytoplasmic retrovirus-like particles, reproducible clearance should be demonstrated in at least two independent studies. Whenever possible, the initial virus load should be determined from the virus which can be detected in the spiked starting material. If this is not possible, the initial virus load may be calculated from the titre of the spiking virus preparation. Where inactivation is too rapid to plot an inactivation curve using process conditions, appropriate controls should be performed to demonstrate that infectivity is indeed lost by inactivation.

Function and regeneration of columns

Over time and after repeated use, the ability of chromatography columns and other devices used in the purification scheme to clear virus may vary. Some estimate of the stability of the viral clearance after several uses may provide support for repeated use of such columns. Assurance should be provided that any virus potentially retained by the production system would be adequately destroyed or removed prior to reuse of the system. For example, such evidence may be provided by demonstrating that the cleaning and regeneration procedures do inactivate or remove virus.

Specific precautions

a) Care should be taken in preparing the high-titer virus to avoid aggregation which may enhance physical removal and decrease inactivation thus distorting the correlation with actual production.

b) Consideration should be given to the minimum quantity of virus which can be reliably assayed.

c) The study should include parallel control assays to assess the loss of infectivity of the virus due to such reasons as the dilution, concentration, filtration or storage of samples before titration.

d) The virus «spike» should be added to the product in a small volume so as not to dilute or change the characteristics of the product. Diluted, test-protein sample is no longer identical to the product obtained at commercial scale.

e) Small differences in, for example, buffers, media, or reagents, can substantially affect viral clearance.

f) Virus inactivation is time-dependent, therefore, the amount of time a spiked product remains in a particular buffer solution or on a particular chromatography column should reflect the conditions of the commercial-scale process.

g) Buffers and product should be evaluated independently for toxicity or interference in assays used to determine the virus titre, as these components may adversely affect the indicator cells. If the solutions are toxic to the indicator cells, dilution, adjustment of the pH, or dialysis of the buffer containing spiked virus might be necessary. If the product itself has anti-viral activity, the clearance study may need to be performed without the product in a «mock» run, although omitting the product or substituting a similar protein that does not have anti-viral activity could affect the behaviour of the virus in some production steps. Sufficient controls to demonstrate the effect of procedures used solely to prepare the sample for assay (e.g., dialysis, storage) on the removal/inactivation of the spiking virus should be included.

h) Many purification schemes use the same or similar buffers or columns repetitively. The effects of this approach should be taken into account when analysing the data. The effectiveness of virus elimination by a particular process may vary with the stage in manufacture at which it is used.

i) Overall reduction factors may be underestimated where production conditions or buffers are too cytotoxic or virucidal and should be discussed on a case-by-case basis. Overall reduction factors may also be overestimated due to inherent limitations or inadequate design of viral clearance studies.

Interpretation of Viral Clearance Studies

Acceptability

The object of assessing virus inactivation/removal is to evaluate and characterise process steps that can be considered to be effective in inactivating/removing viruses and to estimate quantitatively the overall level of virus reduction obtained by the manufacturing process. For virus contaminants, as in Cases B-E, it is important to show that not only is the virus eliminated or inactivated, but that there is excess capacity for viral clearance built into the purification process to assure an appropriate level of safety for the final product. The amount of virus eliminated or inactivated by the production process should be compared to the amount of virus which may be present in unprocessed bulk.

To carry out this comparison, it is important to estimate the amount of virus in the unprocessed bulk. This estimate should be obtained using assays for infectivity or other methods such as transmission electron microscopy (TEM). The entire purification process should be able to eliminate substantially more virus than is estimated to be present in a single-dose-equivalent of unprocessed bulk. See Appendix 4 for calculation of virus reduction factors and Appendix 5 for calculation of estimated particles per dose.

Manufacturers should recognise that clearance mechanisms may differ between virus classes. A combination of factors must be considered when judging the data supporting the effectiveness of virus inactivation removal procedures. These include:

i) The appropriateness of the test viruses used;

ii) The design of the clearance studies;

Appendix 4: Calculation of reduction factors in studies to determine virus clearance.

The virus reduction factor of an individual purification or inactivation step is defined as the $\log_{10}$ of the ratio of the virus load in the pre-purification material and the virus load in the post-purification material which is ready for use in the next step of the process. If the following abbreviations are used:

Starting material:

$$\text{vol } v'; \text{ titer } 10^{a'};$$
$$\text{virus load: } (v')\,(10^{a'}),$$

Final material:

$$\text{vol } v''; \text{ titer } 10^{a''}, \text{ virus load: } (v'')\,(10^{a''}),$$

the individual reduction factors Ri are calculated according to

$$10^{Ri} = (v')(10^{a'}) \,/\, (v'')(10^{a''})$$

This formula takes into account both the titres and volumes of the materials before and after the purification step.

Because of the inherent imprecision of some virus titrations, an individual reduction factor used for the calculation of an overall reduction factor should be greater than 1.

The overall reduction factor for a complete production process is the sum logarithm of the reduction factors of the individual steps. It represents the logarithm of the ratio of the virus load at the beginning of the first process clearance step and at the end of the last process clearance step. Reduction factors are normally expressed on a logarithmic scale which implies that, while residual virus infectivity will never be reduced to zero, it may be greatly reduced mathematically.

Appendix 5: Calculation of estimated particles per dose.

This is applicable to those viruses for which an estimate of starting numbers can be made, such as endogenous retroviruses.

Example:

1. Assumptions

Measured or estimated concentration of virus in cell culture harvest = 10^6/ml
Calculated virus clearance factor = > 10^5
Volume of culture harvest needed to make a dose of product = 1 litre (10^3 ml)

2. Calculation of estimated particles/dose

$$\frac{(10^6 \text{ virus units/ml}) \times (10^3 \text{ ml/dose})}{\text{Clearance factor} > 10^{15}}$$

$$\frac{10^9 \text{ particles/dose}}{\text{Clearance factor} > 10^{15}}$$

$$= < 10^{-6} \text{ particles/dose}$$

Therefore, less than one particle per million doses would be expected.

iii) The log reduction achieved;

iv) The time dependence of inactivation;

v) The potential effects of variation in process parameters on virus inactivation/removal;

vi) The limits of assay sensitivities;

vii)The possible selectivity of inactivation/removal procedure(s) for certain classes of viruses.

Effective clearance may be achieved by any of the following: multiple inactivation steps, multiple complementary separation steps, or combinations of inactivation and separation steps. Since separation methods may be dependent on the extremely specific physicochemical properties of a virus which influence its interaction with gel matrices and precipitation properties, «model» viruses may be separated in a different manner than a target virus. Manufacturing parameters influencing separation should be properly defined and controlled. Differences may originate from changes in surface properties such as glycosylation. However, despite these potential variables, effective removal can be obtained by a combination of complementary separation steps or combinations of inactivation and separation steps. Therefore, well designed separation steps, such as chromatographic procedures, filtration steps and extractions, can be effective virus removal steps provided that they are performed under appropriately controlled conditions. An effective virus removal step should give reproducible reduction of virus load shown by at least two independent studies.

An overall reduction factor is generally expressed as the sum of the individual factors. However, reduction in virus titre of the order of 1 $\log_{10}$ or less would be considered negligible and would be ignored unless justified.

If little reduction of infectivity is achieved by the production process, and the removal of virus is considered to be a major factor in the safety of the product, a specific, additional inactivation/removal step or steps should be introduced. For ail viruses, manufacturers should justify the acceptability of the reduction factors obtained. Results will be evaluated on the basis of the factors listed above.

Limitations of viral clearance studies

Viral clearance studies are useful for contributing to the assurance that an acceptable lever of safety in the final product is achieved but do not by themselves establish safety. However, a number of factors in the design and execution of viral clearance studies may lead to an incorrect estimate of the ability of the process to remove virus infectivity. These factors include the following:

1. Virus preparations used in clearance studies for a production process are likely to be produced in tissue culture. The behaviour of a tissue culture virus in a production step may be different from that of the native virus; for example, if native and cultured viruses differ in purity or degree of aggregation.

2. Inactivation of virus infectivity frequently follows a biphasic curve in which a rapid initial phase is followed by a slower phase. It is possible that virus escaping

a first inactivation step may be more resistant to subsequent steps. For example, if the resistant fraction takes the form of virus aggregates, infectivity may be resistant to a range of different chemical treatments and to heating.

3. The ability of the overall process to remove infectivity is expressed as the sum of the logarithm of the reductions at each step. The summation of the reduction factors of multiple steps, particularly of steps with little reduction (e.g., below 1 $\log_{10}$), may overestimate the true potential for virus elimination. Furthermore, reduction values achieved by repetition of identical or near identical procedures should not be included unless justified.

4. The expression of reduction factors as logarithmic reductions in titre implies that, while residual virus infectivity may be greatly reduced, it will never be reduced to zero. For example, a reduction in the infectivity of a preparation containing 8 $\log_{10}$ infectious units per ml by a factor of 8 $\log_{10}$ leaves zero $\log_{10}$ per ml or one infectious unit per ml, taking into consideration the limit of detection of the assay.

5. Pilot-plant scale processing may differ from commercial-scale processing despite care taken to design the scaled-down process.

6. Addition of individual virus reduction factors resulting from similar inactivation mechanisms along the manufacturing process may overestimate overall viral clearance.

Statistics

The viral clearance studies should include the use of statistical analysis of the data to evaluate the results. The study results should be statistically valid to support the conclusions reached (refer to Appendix 3).

Re-evaluation of viral clearance

Whenever significant changes in the production or purification process are made, the effect of that change, both direct and indirect, on viral clearance should be considered and the system re-evaluated as needed. For example, changes in production processes may cause significant changes in the amount of virus produced by the cell line; changes in process steps may change the extent of viral clearance.

Summary

This document suggests approaches for the evaluation of the risk of viral contamination and for the removal of virus from product, thus contributing to the production of safe biotechnology products derived from animal or human cell lines and emphasises the value of many strategies, including:

A. thorough characterisation/screening of cell substrate starting material in order to identify which, if any, viral contaminants are present;

B. assessment of risk by determination of the human tropism of the contaminants;

C. establishment of an appropriate program of testing for adventitious viruses in unprocessed bulk;

D. careful design of viral clearance studies using different methods of virus inactivation or removal in the same production process in order to achieve maximum viral clearance and

E. performance of studies which assess virus inactivation and removal.

Glossary

Adventitious Virus: See virus.

Cell Substrate: Cells used to manufacture product.

Endogenous Virus: See virus.

Inactivation: Reduction of virus infectivity caused by chemical or physical modification.

In Vitro Cell Age: A measure of the period between thawing of the MCB vial(s) and harvest of the production vessel measured by elapsed chronological time in culture, population doubling level of the cells or passage level of the cells when subcultivated by a defined procedure for dilution of the culture.

Master Cell Bank (MCB): An aliquot of a single pool of cells which generally has been prepared from the selected cell clone under defined conditions, dispensed into multiple containers and stored under defined conditions. The MCB is used to derive all working cell banks. The testing performed on a new MCB (from a previous initial cell clone, MCB or WCB) should be the same as for the MCB, unless justified.

Minimum Exposure Time: The shortest period for which a treatment step will be maintained.

Non-Endogenous Virus: See virus.

Process Characterisation of Viral Clearance: Viral clearance studies in which non-specific «model» viruses are used to assess the robustness of the manufacturing process to remove and/or inactivate viruses.

Process Evaluation Studies of Viral Clearance: Viral clearance studies in which «relevant» and/or specific «model» viruses are used to determine the ability of the manufacturing process to remove and/or inactivate these viruses.

Production Cells: Cell substrate used to manufacture product.

Unprocessed Bulk: One or multiple pooled harvests of cells and culture media. When cells are not readily accessible, the unprocessed bulk would constitute fluid harvested from the fermenter.

Virus: Intracellularly replicating infectious agents that are potentially pathogenic, possessing only a single type of nucleic acid (either RNA or DNA), are unable to grow and undergo binary fission, and multiply in the form of their genetic material.

Adventitious virus: Unintentionally introduced contaminant viruses.

Endogenous virus: Viral entity whose genome is part of the germ line of the species of origin of the cell line and is covalently integrated into the genome of animal from which the parental cell line was derived. For the purposes of this document, intentionally introduced, non-integrated viruses such as EBV used to immortalise cell substrates or Bovine Papilloma Virus fit in this category.

Non-endogenous virus: Viruses from external sources present in the Master Cell Bank.

Non-specific model virus: A virus used for characterisation of viral clearance of the process when the purpose is to characterise the capacity of the manufacturing process to remove and/or inactivate viruses in general, i.e., to characterise the robustness of the purification process.

Relevant virus: Virus used in process evaluation studies which is either the identified virus, or of the same species as the virus that is known, or likely to contaminate the cell substrate or any other reagents or materials used in the production process.

Specific model virus: Virus which is closely related to the known or suspected virus (same genus or family), having similar physical and chemical properties to those of the observed or suspected virus.

Virus Clearance: Elimination of target virus by removal of viral particles or inactivation of viral infectivity.

Virus-like Particles: Structures visible by electron microscopy which morphologically appear to be related to known viruses.

Virus Removal: Physical separation of virus particles from the intended product.

Working Cell Bank (WCB): The WCB is prepared from aliquots of a homogeneous suspension of cells obtained from culturing the MCB under defined culture conditions.

ICH Topic Q5B
Quality of Biotechnological Products: Analysis of the Expression Construct in Cell Lines Used for Production of r-DNA Derived Protein Products

Note for Guidance on Quality of Biotechnological Products: Analysis of the Expression Construct in Cell Lines Used for Production of r-DNA Derived Protein Products (CPMP/ICH/139/95)

Transmission to CPMP	**December 1995**
Final approval by CPMP	**December 1995**
Proposed date for coming into operation (Studies commencing after)	**July 1996**

Brown F, Griffiths E, Horaud F, Petricciani JC (eds): Safety of Biological Products Prepared from Mammalian Cell Culture.
Dev Biol Stand. Basel, Karger, 1998, vol 93, pp 205-208

Quality of Biotechnological Products: Analysis of the Expression Construct in Cell Lines Used for Production of r-DNA Derived Protein Products

ICH Harmonised Tripartite Guideline

Introduction

This document presents guidance regarding the characterisation of the expression construct for the production of recombinant DNA protein products in eukaryotic and prokaryotic cells. This document is intended to describe the types of information that are considered valuable in assessing the structure of the expression construct used to produce recombinant DNA derived proteins. This document is not intended to cover the whole quality aspect of rDNA derived medicinal products.

The expression construct is defined as the expression vector containing the coding sequence of the recombinant protein. Segments of the expression construct should be analysed using nucleic acid techniques in conjunction with other tests performed on the purified recombinant protein for assuring the quality and consistency of the final product. Analysis of the expression construct at the nucleic acid level should be considered as part of the overall evaluation of quality, taking into account that this testing only evaluates the coding sequence of a recombinant gene and not the translational fidelity nor other characteristics of the recombinant protein, such as secondary structure, tertiary structure, and post-translational modifications.

Rationale for Analysis of the Expression Construct

The purpose of analysing the expression construct is to establish that the correct coding sequence of the product has been incorporated into the host cell and is maintained during culture to the end of production. The genetic sequence of recombinant proteins produced in living cells can undergo mutations that could alter the properties of the protein with potential adverse consequences to patients. No single experimental approach can be expected to detect all possible modifications to a protein. Protein analytical techniques can be used to assess the amino acid sequence of the protein and structural features of the expressed protein due to post-translational modifications such as proteolytic processing, glycosylation, phosphorylation, and acetylation. Data from nucleic acid analysis may be useful since protein analytical methods may not detect all changes in protein structure resulting from mutations in the sequence coding for the recombinant protein. The relative importance of nucleic acid analysis and protein analysis will vary from product to product.

Nucleic acid analysis can be used to verify the coding sequence and the physical state of the expression construct. The nucleic acid analysis is performed to ensure that the expressed protein will have the correct amino acid sequence but is not intended to detect low levels of variant sequences. Where the production cells have multiple integrated copies of the expression construct, not all of which may be transcriptionally active, examination of the transcription product itself by analysis of mRNA or cDNA may be more appropriate than analysis of genomic DNA. Analytical approaches that examine a bulk population of nucleic acids, such as those performed on pooled clones or material amplified by the polymerase chain reaction, may be considered as an alternative to approaches that depend on selection of individual DNA clones. Other techniques could be considered that allow for rapid and sensitive confirmation of the sequence coding for the recombinant protein in the expression construct.

The following sections describe information that should be supplied regarding the characterisation of the expression construct during the development and validation of the production system. Analytical methodologies should be validated for the intended purpose of confirmation of sequence. The validation documentation should at a minimum include estimates of the limits of detection for variant sequences. This should be performed for either nucleic acid or protein sequencing methods. The philosophy and recommendations for analysis expressed in this document should be periodically reviewed to take advantage of new advances in technology and scientific information.

Characterisation of the Expression System

Expression construct and cell clone used to develop the Master Cell Bank (MCB)

The manufacturer should describe the origin of the nucleotide sequence coding for the protein. This should include identification and source of the cell from which the nucleotide sequence was originally obtained. Methods used to prepare the DNA coding for the protein should be described.

The steps in the assembly of the expression construct should be described in detail. This description should include the source and function of the component parts of the expression construct, e.g., origins of replication, antibiotic resistance genes, promoters, enhancers, whether or not the protein is being synthesised as a fusion protein. A detailed component map and a complete annotated sequence of the plasmid should be given, indicating those regions that have been sequenced during the construction and those taken from the literature. Other expressed proteins encoded by the plasmid should be indicated. The nucleotide sequence of the coding region of the gene of interest and associated flanking regions that are inserted into the vector, up to and including the junctions of insertion, should be determined by DNA sequencing of the construct.

A description of the method of transfer of the expression construct into the host cell should be provided. In addition, methods used to amplify the expression construct and criteria used to select the cell clone for production should be described in detail.

Cell Bank system

Production of the recombinant protein should be based on well-defined Master and Working Cell Banks. A cell bank is a collection of ampoules of uniform composition stored under defined conditions each containing an aliquot of a single pool of cells. The Master Cell Bank (MCB) is generally derived from the selected cell clone containing the expression construct. The Working Cell Bank (WCB) is derived by expansion of one or more ampoules of the MCB. The cell fine history and production of the cell banks should be described in detail, including methods and reagents used during culture, in-vitro cell age, and storage conditions. All cell banks should be characterised for relevant phenotypic and genotypic markers which could include the expression of the recombinant protein or presence of the expression construct.

The expression construct in the MCB should be analysed as described below. If the testing cannot be carried out on the MCB, it should be carried out on each WCB.

Restriction endonuclease mapping or other suitable techniques should be used to analyse the expression construct for copy number, for insertions or deletions, and for the number of integration sites. For extrachromosomal expression systems, the percent of host cells retaining the expression construct should be determined.

The protein coding sequence for the recombinant protein product of the expression construct should be verified. For extrachromosomal expression systems, the expression construct should be isolated and the nucleotide sequence encoding the product should be verified without further cloning. For cells with chromosomal copies of the expression construct, the nucleotide sequence encoding the product could be verified by recloning and sequencing of chromosomal copies. Alternatively, the nucleic acid sequence encoding the product could be verified by techniques such as sequencing of pooled cDNA clones or material amplified by the polymerase chain reaction. The nucleic acid sequence should be identical, within the limits of detection of the methodology, to that determined for the expression construct as described in Section III.A. and should correspond to that expected for the protein sequence.

Limit for in-vitro cell age for production

The limit for in-vitro cell age for production should be based on data derived from production cells expanded under pilot or full scale conditions to the proposed in-vitro cell age or beyond. Generally, the production cells are obtained by expansion of the Working Cell Bank; the Master Cell Bank could be used to prepare the production cells with appropriate justification.

The expression construct of the production cells should be analysed once for the MCB as described in Section III.B, except that the protein coding sequence of the expression construct in the production cells could be verified by either nucleic acid testing or analysis of the final protein product. Increases in the defined limit for in-vitro cell age for production should be supported by data from cells which have been expanded to an in-vitro cell age which is equal to or greater than the new limit for in-vitro cell age.

Conclusion

The characterisation of the expression construct and the final purified protein are both important to ensure the consistent production of a recombinant DNA derived product. As described above, it is considered that analytical data derived from both nucleic acid analysis and evaluation of the final purified protein should be evaluated to ensure the quality of a recombinant protein product.

Glossary of Terms

Expression Construct: The expression vector which contains the coding sequence of the recombinant protein and the elements necessary for its expression.

Flanking Control Regions: Non-coding nucleotide sequences that are adjacent to the 5' and 3' end of the coding sequence of the product which contain important elements that affect the transcription, translation, or stability of the coding sequence. These regions include, e.g. promoter, enhancer, and splicing sequences and do not include origins of replication and antibiotic resistance genes.

Integration Site: The site where one or more copies of the expression construct is integrated into the host cell genome.

In-vitro Cell Age: Measure of time between thaw of the MCB vial(s) to harvest of the production vessel measured by elapsed chronological time in culture, by population doubling level of the cells, or by passage level of the cells when subcultivated by a defined procedure for dilution of the culture.

Master Cell Bank (MCB): An aliquot of a single pool of cells which generally has been prepared from the selected cell clone under defined conditions, dispensed into multiple containers and stored under defined conditions. The MCB is used to derive all working cell banks. The testing performed on a new MCB (from a previous initial cell clone, MCB or WCB) should be the same as for the MCB unless justified.

Pilot Plant Scale: The production of a recombinant protein by a procedure fully representative of and simulating that to be applied on a full commercial manufacturing scale. The methods of cell expansion, harvest, and product purification should be identical except for the scale of production.

Relevant Genotypic and Phenotypic Markers: Those markers permitting the identification of the strain of the cell line which should include the expression of the recombinant protein or presence of the expression construct.

Working Cell Bank (WCB): The Working Cell Bank is prepared from aliquots of a homogeneous suspension of cells obtained from culturing the MCB under defined culture conditions.

ICH Topic Q5C
Quality of Biotechnological Products: Stability Testing of Biotechnological/ Biological Products

Note for Guidance on Quality of Biotechnological Products: Stability Testing of Biotechnological/ Biological Products (CPMP/ICH/138/95)

Transmission to CPMP	**December 1995**
Final approval by CPMP	**December 1995**
Proposed date for coming into operation (Studies commencing after)	**July 1996**

Brown F, Griffiths E, Horaud F, Petricciani JC (eds): Safety of Biological Products Prepared from Mammalian Cell Culture.
Dev Biol Stand. Basel, Karger, 1998, vol 93, pp 211-219

Quality of Biotechnological Products: Stability Testing of Biotechnological/ Biological Products

Annex to the ICH Harmonised Tripartite Guideline for the Stability Testing of New Drug Substances and Products

PREAMBLE

The guidance stated in the ICH Harmonised Tripartite Guideline «Stability Testing of New Drug Substances and Products» (27 October 1993) applies in general to biotechnological/biological products. However, biotechnological/biological products do have distinguishing characteristics to which consideration should be given in any well-defined testing program designed to confirm their stability during the intended storage period. For such products, in which the active components are typically proteins and/or polypeptides, maintenance of molecular conformation and, hence of biological activity, is dependent on noncovalent as well as covalent forces. The products are particularly sensitive to environmental factors such as temperature changes, oxidation, light, ionic content, and shear. In order to ensure maintenance of biological activity and to avoid degradation, stringent conditions for their storage are usually necessary.

The evaluation of stability may necessitate complex analytical methodologies. Assays for biological activity, where applicable, should be part of the pivotal stability studies. Appropriate physicochemical, biochemical and immunochemical methods for the analysis of the molecular entity and the quantitative detection of degradation products should also be part of the stability program whenever purity and molecular characteristics of the product permit use of these methodologies.

With the above concerns in mind, the applicant should develop the proper supporting stability data for a biotechnological/biological product and consider many external conditions which can affect the product's potency, purity and quality. Primary data to support a requested storage period for either drug substance or drug product should always be based on long-term, real-time, real-condition stability studies. Thus, the development of a proper long-term stability program becomes critical to the successful development of a commercial product. The purpose of this document is to give guidance to applicants regarding the type of stability studies that should be provided in support of marketing applications. It is understood that during the review and evaluation process, continuing updates of initial stability data may occur.

Scope of the Annex

The guidance stated in this annex applies to well-characterised proteins and polypeptides, their derivatives and products of which they are components, and which are isolated from «issues, body fluids, cell cultures, or produced using rDNA technology. Thus, the document covers the generation and submission of stability data for products such as cytokines (interferons, interleukins, colony-stimulating factors, tumour necrosis factors), erythropoietins, plasminogen activators, blood plasma factors, growth hormones and growth factors, insulins, monoclonal antibodies, and vaccines consisting of well-characterised proteins or polypeptides. In addition, the guidance outlined in the following sections may apply to other types of products, such as conventional vaccines, after consultation with the appropriate regulatory authorities. The document does not cover antibiotics, allergenic extracts, heparins, vitamine or whole blood.

Terminology

For the basic terms used in this annex the reader is referred to the «Glossary» in the ICH~Harmonised Tripartite Guideline «Stability Testing of New Drug Substances and Products» (27 October 1993). However, since manufacturers of biotechnological/biological products sometimes use traditional terminology, traditional terms are specified in parentheses to assist the reader. A supplemental glossary is also included that explains certain terms used in the production of biotechnological/biological products.

Selection of Batches

Drug substance (Bulk material)

Where bulk material is to be stored after manufacture but prior to formulation and final manufacturing, stability data should be provided on at least three batches for which manufacture and storage are representative of the manufacturing scale of production. A minimum of six months stability data at the time of submission should be submitted in cases where storage periods greater than six months are requested. For drug substances with storage periods of less than six months, the minimum amount of stability data in the initial submission should be determined on a case-by-case basis. Data from pilot-plant-scale batches of drug substance produced at a reduced scale of fermentation and purification may be provided at the time the dossier is submitted to the regulatory agencies with a commitment to place the first three manufacturing scale batches into the long-term stability program after approval.

The quality of the batches of drug substance placed into the stability program should be representative of the quality of the material used in pre-clinical and clinical studies and of the quality of the material to be made at manufacturing scale. In addition, the drug substance (bulk material) made at pilot-plant scale should be produced by a process and stored under conditions representative of that used for the manufacturing scale. The drug substance entered into the stability program should be stored in containers which properly represent the actual holding containers used during manufacture. Containers of reduced size may be acceptable for drug substance stability testing provided that they are constructed of the same material and use the same type of container/closure system that is intended to be used during manufacture.

Intermediates

During manufacture of biotechnological/biological products, the quality and control of certain intermediates may be critical to the production of the final product. In general, the manufacturer should identify intermediates and generate in-house data and process limits that assure their stability within the bounds of the developed process. While the use of pilot-plant-scale data is permissible, the manufacturer should establish the suitability of such data using the manufacturing-scale process.

Drug product (final container product)

Stability information should be provided on at least three batches of final container product representative of that which will be used at manufacturing scale. Where possible, batches of final container product included in stability testing should be derived from different batches of bulk material. A minimum of six months data at the time of submission should be submitted in cases where storage periods greater than six months are requested. For drug products with storage periods of less than six months, the minimum amount of stability data in the initial submission should be determined on a case-by-case basis. Product expiration dating will be based upon the actual data submitted in support of the application. Since dating is based upon the real-time/real-temperature data submitted for review, continuing updates of initial stability data should occur during the review and evaluation process. The quality of the final container product placed on stability studies should be representative of the quality of the material used in the preclinical and clinical studies. Data from pilot-plant scale batches of drug product may be provided at the time the dossier is submitted to the regulatory agencies with a commitment to place the first three manufacturing scale batches into the long-term stability program after approval. Where pilot-plant scale batches were submitted to establish the dating for a product and, in the event that product produced at manufacturing scale does not meet those long-term stability specifications throughout the dating period or is not representative of the material used in pre-clinical and clinical studies, the applicant should notify the appropriate regulatory authorities to determine a suitable course of action.

Sample selection criteria

Where one product is distributed in batches differing in fill volume (e.g., 1 milliliter (ml), 2 ml, or 10 ml), unitage (e.g., 10 units, 20 units, or 50 units), or mass (e.g., 1 milligram (mg), 2 mg, or 5 mg) samples to be entered into the stability program may be selected on the basis of a matrix system and/or by bracketing.

Matrixing, i.e., the statistical design of a stability study in which different fractions of samples are tested at different sampling points, should only be applied when appropriate documentation is provided that confirms that the stability of the samples tested represents the stability of all samples. The differences in the samples for the same drug product should be identified as, for example, covering different batches, different strengths, different sizes of the same closure and possibly, in some cases, different container/closure systems. Matrixing should not be applied to samples with differences that may affect stability, such as different strengths and different containers/closures, where it cannot be confirmed that the products respond similarly under storage conditions.

Where the same strength and exact container/closure system is used for three or more fill contents, the manufacturer may elect to place only the smallest and largest container size into the stability program, i.e., bracketing. The design of a protocol that incorporates bracketing assumes that the stability of the intermediate condition samples are represented by those at the extremes. In certain cases, data may be needed to demonstrate that all samples are properly represented by data collected for the extremes.

Stability-Indicating Profile

On the whole, there is no single stability-indicating assay or parameter that profiles the stability characteristics of a biotechnological/biological product. Consequently, the manufacturer should propose a stability-indicating profile that provides assurance that changes in the identity, purity and potency of the product will be detected.

At the time of submission, applicants should have validated the methods that comprise the stability-indicating profile and the data should be available for review. The determination of which tests should be included will be product-specific. The items emphasised in the following subsections are not intended to be all-inclusive, but represent product characteristics that should typically be documented to adequately demonstrate product stability.

Protocol

The dossier accompanying the application for marketing authorisation should include a detailed protocol for the assessment of the stability of both drug substance and drug product – in support of the proposed storage conditions and expiration dating periods. The protocol should include all necessary information which demonstrates the stability of the biotechnological/biological product throughout the proposed expiration dating period including, for example, well-defined specifications and test intervals. The statistical methods that should be used are described in the Tripartite Guideline on stability.

Potency

When the intended use of a product is linked to a definable and measurable biological activity, testing for potency should be part of the stability studies. For the purpose of stability testing of the products described in this guideline, potency is the specific ability or capacity of a product to achieve its intended effect. It is based on the measurement of some attribute of the product and is determined by a suitable quantitative method. In general, potencies of biotechnological/biological products tested by different laboratories can be compared in a meaningful way only if expressed in relation to that of an appropriate reference material. For that purpose, a reference material calibrated directly or indirectly against the corresponding national or international reference material should be included in the assay.

Potency studies should be performed at appropriate intervals as defined in the stability protocol and the results should be reported in units of biological activity calibrated, whenever possible, against nationally or internationally recognised standard. Where no national or international standards exists, the assay results may be reported in in-house derived units using a characterised reference material.

In some biotechnological/biological products, potency is dependent upon the conjugation of the active ingredient(s) to a second moiety or binding to an adjuvant. Dissociation of the active ingredient(s) from the carrier used in conjugates or adjuvants should be examined in real-time/real-temperature studies (including conditions encountered during shipment). The assessment of the stability of such products may be difficult since, in some cases, in vitro tests for biological activity and physicochemical characterisation are impractical or provide accurate results. Appropriate strategies (e.g., testing the product prior to conjugation/binding, assessing the release of the active compound from the second moiety, in vivo assays) or the use of an appropriate surrogate test should be considered to overcome the inadequacies of in vitro testing.

Purity and molecular characterisation

For the purpose of stability testing of the products described in this guideline, purity is a relative term. Due to the effect of glycosylation, deamidation, or other heterogeneities, the absolute purity of a biotechnological/biological product is extremely difficult to determine. Thus, the purity of a biotechnological/biological product should be typically assessed by more than one method and the purity value derived is method-dependent. For the purpose of stability testing, tests for purity should focus on methods for determination of degradation products.

The degree of purity, as well as individual and total amounts of degradation products of the biotechnological/biological product entered into the stability studies, should be reported and documented whenever possible. Limits of acceptable degradation should be derived from the analytical profiles of batches of the drug substance and drug product used in the pre-clinical and clinical studies.

The use of relevant physicochemical, biochemical and immunochemical analytical methodologies should permit a comprehensive characterisation of the drug substance and/or drug product (e.g., molecular size, charge, hydrophobicity) and the accurate detection of degradation changes that may result from deamidation, oxidation, sulfoxidation, aggregation or fragmentation during storage. As examples, methods that may contribute to this include electrophoresis (SDS-PAGE, immunoelectrophoresis, Western blot, isoelectrofocusing), high-resolution chromatography (e.g., reversed-phase chromatography, gel filtration, ion exchange, affinity chromatography), and peptide mapping.

Wherever significant qualitative or quantitative changes indicative of degradation product formation are detected during long-term, accelerated and/or stress stability studies, consideration should be given to potential hazards and to the need for characterisation and quantification of degradation products within the long-term stability program. Acceptable limits should be proposed and justified, taking into account the levels observed in material used in pre-clinical and clinical studies.

For substances that can not be properly characterised or products for which an exact analysis of the purity cannot be meaningfully determined through routine analytical methods, the applicant should propose and justify alternative testing procedures.

Other product characteristics

The following product characteristics, though not specifically relating to biotechnological/biological products, should be monitored and reported for the drug product in its final container:

- Visual appearance of the product (colour and opacity for solutions/suspensions; colour, texture and dissolution time for powders), visible particulates in solutions or after the reconstitution of powders or lyophilised cakes, pH, and moisture level of powders and lyophilised products.

- Sterility testing or alternatives (e.g., container/closure integrity testing) should be performed at a minimum initially and at the end of the proposed shelf-life.

- Additives (e.g., stabilisers, preservatives) or excipients may degrade during the dating period of the drug product. If there is any indication during preliminary stability studies that reaction or degradation of such materials adversely affect the quality of the drug product, these items may need to be monitored during the stability program.

- The container/closure has the potential to adversely affect the product and should be carefully evaluated (see below).

Storage Conditions

Temperature

Since most finished biotechnological/biological products need precisely defined storage temperatures, the storage conditions for the real-time/real-temperature stability studies may be confined to the proposed storage temperature.

Humidity

Biotechnological/biological products are generally distributed in containers protecting them against humidity. Therefore, where it can be demonstrated that the proposed containers (and conditions of storage) afford sufficient protection against high and low humidity, stability tests at different relative humidities can usually be omitted. Where humidity-protecting containers are not used, appropriate stability data should be provided.

Accelerated and stress conditions

As previously noted, the expiration dating should be based on real-time/real-temperature data. However, it is strongly suggested that studies be conducted on the drug substance and drug product under accelerated and stress conditions. Studies under accelerated conditions may provide useful support data for establishing the expiration date, provide product stability information for future product development (e.g., preliminary assessment of proposed manufacturing changes such as change in formulation, scale-up), assist in validation of analytical methods for the stability program, or generate information which may help elucidate the degradation profile of the drug substance or drug product. Studies under stress conditions may be useful in determining whether accidental exposures to conditions other than those proposed (e.g., during transportation) are deleterious to the product and also for evaluating which specific test parameters may be the best indicators of product stability. Studies of the exposure of the drug substance or drug product to extreme conditions may help to reveal patterns of degradation; if so, such changes should be monitored under proposed storage conditions. While the Tripartite Guideline on stability describes the conditions of the accelerated and stress study, the applicant should note that those conditions may not be appropriate for biotechnological/biological products. Conditions should be carefully selected on a case-by-case basis.

Light

Applicants should consult the appropriate regulatory authorities on a case-by-case basis to determine guidance for testing.

Container/Closure

Changes in the quality of the product may occur due to the interactions between the formulated biotechnological/biological product and container/closure. Where the lack of interactions cannot be excluded in liquid products (other than sealed ampoules), stability studies should include samples maintained in the inverted or horizontal position (i.e., in contact with the closure), as well as in the upright position, to determine the effects of the closure on product quality. Data should be supplied for all different container/closure combinations that will be marketed.

In addition to the standard data necessary for a conventional single-use vial, the applicant should demonstrate that the closure used with a multiple-dose vial is capable of withstanding the conditions of repeated insertions and withdrawals so that the product retains its full potency, purity, and quality for the maximum period specified in the instructions-for-use on containers, packages, and/or package inserts. Such labelling should be in accordance with relevant national/regional requirements.

Stability after reconstitution of freeze-dried product

The stability of freeze-dried products after their reconstitution should be demonstrated for the conditions and the maximum storage period specified on containers, packages, and/or package inserts. Such labelling should be in accordance with relevant national/regional requirements.

Testing Frequency

The shelf-lives of biotechnological/biological products may vary from days to several years. Thus, it is difficult to draft uniform guidelines regarding the stability study duration and testing frequency that would be applicable to all types of biotechnological/biological products. With only a few exceptions, however, the shelf-lives for existing products and potential future products will be within the range of 0.5 to five years. Therefore, the guidance is based upon expected shelf-lives in that range. This takes into account the fact that degradation of biotechnological/biological products may not be governed by the same factors during different intervals of a long storage period.

When shelf-lives of one year or less are proposed, the real-time stability studies should be conducted monthly for the first three months and at three-month intervals thereafter.

For products with proposed shelf-lives of greater than one year, the studies should be conducted every three months during the first year of storage, every six months during the second year, and annually thereafter.

While the testing intervals listed above may be appropriate in the pre-approval or pre-license stage, reduced testing may be appropriate after approval or licensure where data are available that demonstrate adequate stability. Where data exist that indicate the stability of a product is not compromised, the applicant is encouraged to submit a protocol which supports elimination of specific test intervals (e.g., nine-month testing) for post-approval/post-licensure, long-term studies.

Specifications

Although biotechnological/biological products may be subject to significant losses of activity, physicochemical changes, or degradation during storage, international and national regulations have provided little guidance with respect to distinct release and end of shelf-life specifications. Recommendations for maximum acceptable losses of activity, limits for physicochemical changes, or degradation during the proposed shelf-life have not been developed for individual types or groups of biotechnological/biological products but are considered on a case-by-case basis. Each product should retain its specifications within established limits for safety, purity, and potency throughout its proposed shelf-life. These specifications and limits should be derived from all available information using the appropriate statistical methods. The use of different specifications for release and expiration should be supported by sufficient data to demonstrate that clinical performance is not affected as discussed in the Tripartite Guideline on stability.

Labelling

For most biotechnological/biological drug substances and drug products, precisely defined storage temperatures are recommended. Specific recommendations should be stated, particularly for drug substances and drug products that cannot tolerate freezing. These conditions, and where appropriate, recommendations for protection against light and/or humidity, should appear on containers, packages, and/or package inserts. Such labelling should be in accordance with relevant national/regional requirements.

Glossary

Conjugated Product: A conjugated product is made up of an active ingredient (for example, peptide, carbohydrate) bound covalently or noncovalentlv to a carrier (for example, protein, peptide, inorganic mineral) with the objective of improving the efficacy or stability of the product.

Degradation Product: A molecule resulting from a change in the drug substance (bulk material) brought about over time. For the purpose of stability testing of the products described in this guideline, such changes could occur as a result of processing or storage (e.g., by deamidation, oxidation, aggregation, proteolysis). For biotechnological/biological products some degradation products may be active.

Impurity: Any component of the drug substance (bulk material) or drug product (final container product) which is not the chemical entity defined as the drug substance, an excipient, or other additives to the drug product.

Intermediate: For biotechnological/biological products, a material produced during a manufacturing process which is not the drug substance or the drug product but whose manufacture is critical to the successful production of the drug substance or the drug product. Generally, an intermediate will be quantifiable and specifications will be established to determine the successful completion of the manufacturing step prior to continuation of the manufacturing process. This includes material which may undergo further molecular modification or be held for an extended period of time prior to further processing.

Manufacturing-Scale Production: Manufacture at the scale typically encountered in a facility intended for product production for marketing.

Pilot-Plant Scale: The production of the drug substance or drug product by a procedure fully representative of and simulating that to be applied at manufacturing scale. The methods of cell expansion, harvest, and product purification should be identical except for the scale of production.

ICH Topic Q5D
Quality of Biotechnological Products: Derivation and Characterisation of Cell Substrates Used for Production of Biotechnological/Biological Products

Note for Guidance on Quality of Biotechnological Products: Derivation and Characterisation of Cell Substrates Used for Production of Biotechnological/ Biological Products (CPMP/ICH/294/95)

Transmission to CPMP	January 1997
Transmission to interested parties	January 1997
Comments requested before	July 1997
Final approval by CPMP	September 1997
Date for coming into operation	March 1998

Brown F, Griffiths E, Horaud F, Petricciani JC (eds): Safety of Biological Products Prepared from Mammalian Cell Culture.
Dev Biol Stand. Basel, Karger, 1998, vol 93, pp 223-234

Quality of Biotechnological Products: Derivation and Characterisation of Cell Substrates Used for Production of Biotechnological/Biological Products

ICH Harmonised Tripartite Guideline

INTRODUCTION

Objective

The objective of this guideline is to provide broad guidance on appropriate standards for the derivation of human and animal cell lines and microbial cells to be used to prepare biotechnological/biological products defined in Section 1.3, Scope, and for the preparation and characterisation of cell banks to be used for production. The document, therefore, provides recommendations on the information in these areas that should be presented in marketing applications for these products.

Rationale

Historically, some quality concerns for cell-derived biological products have originated from the presence of adventitious contaminants or from the properties of the cells used to prepare the product. Recombinant DNA (rDNA)-derived products also carry quality concerns regarding the expression construct contained in the cell substrate. Thus, it is well established that the properties of the cell substrate and events linked to the cell substrate can affect resultant product quality and safety and, further, that effective quality control of these products requires appropriate controls on all aspects of handling the cell substrate.

This document complements other guidelines to provide a comprehensive approach to quality issues arising from biological aspects of processing products from metazoan and microbial cell culture.

Scope

This guideline covers cell substrates having a cell banking system. In this document, «cell substrate» refers to microbial cells or cell lines derived from human or animal sources that possess the full potential for generation of the desired biotechnological/biological products for human in vivo or ex vivo use. Reagents for in vitro diagnostic use are outside the scope of this document. Animal sources of cell lines include all those of metazoan origin. Both continuous cell lines of indefinite in vitro lifespan and diploid cells of finite in vitro lifespan are included. Microbial sources include bacteria, fungi, yeast, and other unicellular life forms.

«Biotechnological/biological products» refers to any products prepared from cells cultivated from cell banks with the exception of microbial metabolites such as, for example, antibiotics, amino acids, carbohydrates, and other low molecular weight substances. Cell banks used to prepare gene therapy products or vaccines should follow the recommendations presented in this document. Some biological products, such as certain viral vaccines, are prepared in primary cell cultures derived directly from animal tissues or organs. Primary cells are not banked and therefore are not addressed by this document. However, other considerations which may apply to primary cells are discussed further in Appendix 1 of this document.

GUIDELINES

Source, history, and generation of the cell substrate

Introduction

It is important to provide supportive documentation which describes the history of the cell substrate that is used in the manufacture of a biotechnological/biological product, as well as any parental cell line from which it was totally or partially derived. Events during the research and development phases of the cell substrate may contribute significantly to assessment of the risks associated with the use of that particular cell substrate for production. The information supplied in this regard is meant to facilitate an overall evaluation which will ensure the quality and safety of the product.

Careful records of the manipulation of the cell substrate should be maintained throughout its development. Description of cell history is only one tool of many used for cell substrate characterisation. In general, deficiencies in documented history may not, by itself, be an impediment to product approval, but extensive deficiencies will result in increased reliance on other methods to characterise the cell substrate.

Origin, source, and history of cells

The source of cells (laboratory or culture collection) from which the cell substrate was derived should be stated, and relevant references from the scientific literature should be cited. Information obtained directly from the source laboratory is preferred. When this is not available, literature references may be utilised.

For human cell lines, it is relevant to describe the following characteristics of the original donor: Tissue or organ of origin, ethnic and geographical origin, age, sex and general physiological condition. If known, the state of health or medical history of the donor should be reported along with the results of any tests of the donor for pathogenic agents. Specifically for human diploid fibroblasts, the age of the donor may influence the in vitro lifespan of the cell line and this information should be provided if available. For animal cell lines, relevant descriptions of the source include species, strains, breeding conditions, tissue or organ of origin, geographical origin, age and sex, the results of tests for pathogenic agents, and general physiological condition of the original donor.

For microbes, manufacturers should describe the species, strain, and known genotypic and phenotypic characteristics of the organism from which the cell substrate was derived. Manufacturers should also describe the pathogenicity, toxin production, and other biohazard information, if any.

The cultivation history of the cells should be documented. The method originally used for the isolation of the cells should be described as well as the procedures used in the culturing of the cells in vitro and any procedures used to establish cell lines (for example, use of any physical, chemical, or biological procedure, or added nucleotide sequences). A description of any genetic manipulation or selection should be provided. All available information regarding the identification, characteristics, and results of testing of these cells for endogenous and adventitious agents should be provided.

For continuous cell lines of metazoan origin, it is usually adequate to quantitate culture duration by estimation of either number of population doublings, or number of subcultivations at defined dilution ratio, or time in days. For diploid cell lines possessing finite in vitro lifespan, accurate estimation of the number of population doublings during all stages of research, development, and manufacturing is important. For microbial cells, documentation of subcultivation frequency after cell substrate generation is considered adequate.

Regarding the generation of cell substrates, applicants should provide a thorough discussion of procedures which would provide exposure to infectious agents. Constituents of the culture medium should be described, in particular, information regarding exposure of the cells to materials of human or animal origin such as serum, enzymes, hydrolysates, or other living cells. The description should include the source, method of preparation and control, test results, and quality assurance. Relevant literature on these points may be referenced when available. This information will allow a detailed analysis of potential entry routes for adventitious agents from these sources, and will be part of the risk-benefit analysis of the product.

Generation of the cell substrate

A crucial step is the choice of a suitable parental cell line. For recombinant products, a parental cell line is typically the untransfected recipient cell line. The use of characterised parental cell banks is suggested, but is not considered essential. A characterised parental cell bank may be of benefit, especially when multiple cell substrates are generated from the same parental cell type, by providing a set of information on which the quality assessment of the Master Cell Bank (MCB) can be based . For example, the myeloma cell line may be banked as a parental cell line for hybridomas.

During the generation of the cell substrate, one or more specific procedures may be utilised in the ultimate development of the desired characteristics. These may include, for example, cell fusion, transfection, selection, colony isolation, cloning, gene amplification, and adaptation to specific culture conditions or media. Information regarding the methodologies utilised in developing the cell substrate can help to provide a clear understanding of the history of the cell substrate. Some cell substrates such as human diploid fibroblasts may not need extensive manipulation or cloning prior to cell banking.

For recombinant products, the cell substrate is the transfected cell containing the desired sequences, which has been cloned from a single cell progenitor. For further information on generation of rDNA-modified cell substrates, consult other relevant (e.g., regional or international) guidelines. For non-recombinant products or non-recombinant vaccines, the cell substrate is the cell from the parental cell line chosen for preparation of the MCB without further modification. For products derived from hybridomas, the cell substrate is the hybridoma cell line derived by fusion of the parental myeloma cell line with other parental cells, e.g., immune spleen cells.

Cell banking

One of the most important advantages of using serially subcultivated cells to produce biotechnological/biological products is the ability to have a characterised common starting source for each production lot, i.e., the preserved bank of cells. Manufacturers may prepare their own cell banks, or may obtain them from external sources. Manufacturers are responsible for ensuring the quality of each cell bank and of the testing performed on each bank.

Cell banking system

The concept of a two-tiered cell bank, in which the MCB which is used to generate Working Cell Banks (WCBs), is generally accepted as the most practical approach to providing a supply of cell substrate for continued manufacture of the product. Manufacturers should describe their strategy for providing a continued supply of cells from their cell bank(s), including the anticipated utilisation rate of the cell bank(s) for production, the expected intervals between generation of new cell bank(s), and the criteria for qualification of cell bank(s).

Generally, the MCB is made first, usually directly from an initial clone or from a preliminary cell bank derived from an initial clone. It is not considered necessary to prepare cell banks from clones for certain types of cells (e.g., diploid cells, where limited in vitro life span or other technical factors make cell cloning impractical) or where the uncloned cell population is already adequately homogeneous for the intended use.

A WCB is derived from one or more containers of the MCB. It is the WCB which is typically used to directly provide cells for the manufacturing process. Additional WCBs are generated from the MCB as needed. A newly prepared WCB should be appropriately qualified by characterisation and testing.

It should be noted that the MCB and WCB may differ from each other in certain respects, e.g., culture components and culture conditions. Similarly, the culture conditions used to prepare the MCB and WCB may differ from those used for the production process. If changes in cell culture process do not affect product quality, it is not considered necessary to reclone the cells or to rebank the MCB or WCB. It is important that a characterised bank provides a consistent product.

A single-tiered banking system consisting only of the MCB but no WCBs could be used in principle, for example, if relatively few containers were needed each year to produce the desired product.

In some microbial expression systems, a new transformation is performed for each new cell substrate container lot, based upon using aliquots of thoroughly tested host cell banks and plasmid banks for each new transformation and on testing of each transformed cell substrate bank. This transformed cell substrate bank is considered the MCB, and it is used as the source of cell substrate for production. Host cell banks, plasmid banks, and MCBs are maintained by appropriate preservation methods. This alternative system is considered adequate because the transformation of bacteria and yeast is generally a very reproducible and easily performed process, unlike the events needed for transfection of metazoan cells . Manufacturers should provide information on the host cells, rDNA molecules (such as plasmids), method of transformation and of cell banking, and the results of characterisation studies.

Cell banking procedures

It is important to prevent a contaminated cell substrate (or bank) from being used in production and to avoid a loss of product availability or development time resulting from the need to recreate a cell bank found to be unusable due to contamination. It is recognised that no cell bank testing regimen is able to detect all potentiel contaminants; therefore, use of these preventive principles during cell banking is important to provide reasonable assurance of the absence of contamination and to provide a reliable source of the cell substrate.

Manufacturers should describe the type of banking system used, the size of the cell bank(s), the container (vials, ampoules, or other appropriate vessels) and closure system used, the methods used for preparation of the cell bank(s) including the cryoprotectants and media used, and the conditions employed for cryopreservation and storage.

Manufacturers should describe the procedures used to avoid microbial contamination and cross-contamination by other cell types present in the laboratory, and the procedures that allow the cell bank containers to be traced. This should include a description of the documentation system as well as that of a labelling system which can withstand the process of preservation, storage, and recovery from storage without loss of labelling information on the container.

Manufacturers should describe their cell banking procedures. Cells are generally prepared for banking by expanding cultures in a progressively greater number or larger size of vessel until a pool of cells can be obtained which is sufficient to generate enough containers for the bank. To ensure the uniform composition of the contents of each container, a single pool of cells for banking should be prepared by combining the cells from all of the culture vessels, if more than one vessel is used.

Cells suspended in preservation medium are aliquoted from the single pool into sterilised containers which are then sealed and stored under appropriate conditions. For example, animal cells in media containing a cryoprotectant are frozen in the sealed containers under defined and controlled conditions, and then transferred to storage in the vapor or liquid phase of liquid nitrogen or at equivalent ultra low temperatures. Other methods of preservation and storage may be adequate depending on the organism used, but they should be capable of maintaining a level of cell viability upon reconstitution which is both consistent and adequate for production use.

To ensure continuous, uninterrupted production of pharmaceuticals, manufacturers should carefully consider the steps that can be taken to provide for protection from catastrophic events that could render the cell bank unusable. Examples of these events include fires, power outages and human error. Manufacturers should describe their plans for such precautions; for example, these may include redundancy in the storage of bank containers in multiple freezers, use of back-up power, use of automatic liquid nitrogen fill systems for storage units, storage of a portion of the MCB and WCB at remote sites, or regeneration of the MCB.

The starting point of reference for estimates of in vitro cell age during manufacturing should be the thawing of one or more containers of the MCB. For diploid cell lines, in vitro lifespan should be estimated in terms of population doubling levels. The population doubling level at which senescence occurs should be determined for diploid cells.

General principles of characterisation and testing of cell banks

The characterisation and testing of banked cell substrates is a critical component of the control of biotechnological and biological products. Characterisation of the MCB allows the manufacturer to assess this source with regard to presence of cells from other lines, adventitious agents, endogenous agents and molecular contaminants (e.g., toxins or antibiotics from the host organism). The objective of this testing is to confirm the identity, purity, and suitability of the cell substrate for manufacturing use. In some cases, additional testing such as tumorigenicity or karyology may be useful. The testing program chosen for a given cell substrate will vary according to the biological properties of the cells (for example, growth requirements), its cultivation history (including use of human-derived and animal-derived biological reagents) and available testing procedures. The extent of characterisation of a cell substrate may influence the type or level of routine testing needed at later stages of manufacturing. Manufacturers should perform tests for identity and purity once for each MCB, and tests of stability during cell cultivation once for each product to be registered. In addition, tests of purity and limited tests of identity should be performed once on each WCB. Also, applicants should consult the ICH guideline on viral safety. Relevant tests among those described below should be performed and described in the market application, along with the results of the testing.

For cell lines containing exogenously assembled expression constructs, the relevant ICH guideline on rDNA expression constructs should be consulted for guidance on the characterisation of nucleotide and amino acid sequences. It may also be useful to examine, by similar methods, the coding sequences in some non-recombinant DNA-derived cell lines where the gene sequences have been characterised and are well understood. However, it is not considered necessary to carry out investigations of the sequences encoding complex naturel products, for example, families of related gene products, microbial vaccine antigens, or monoclonal antibodies from hybridomas.

Manufacturers are also encouraged to employ «state-of-the-art» methods and technological improvements in cell substrate characterisation and testing as they become available, as long as the specificity, sensitivity, and precision of the newer methods are at least equivalent to those of existing methods.

The manufacturer may choose to characterise the WCB instead of the MCB, if justified.

Tests of identity

Appropriate tests should be performed to determine that the banked cell is what it is represented to be. Either phenotypic or genotypic characteristics may be used in identity testing. It is not considered necessary to do all the possible tests. Tests of identity are generally performed on the MCB. In addition, limited identity testing is generally performed on each WCB.

Metazoan cells: For human or animal cells which grow attached to a substratum, morphological analysis may be a useful tool in conjunction with other tests. In most cases, isoenzyme analysis is sufficient to confirm the species of origin for cell lines derived from human or animal sources; other tests may be appropriate depending on the history of the cell line. Other technologies may be substituted to confirm species of origin, including, for example, banding cytogenetics or use of species-specific antisera. An alternative strategy would be to demonstrate the presence of unique markers, for example, by using banding cytogenetics to detect a unique marker chromosome, or DNA analysis to detect a genomic polymorphism pattern (for example, restriction fragment length polymorphism, variable number of tandem repeats, or genomic dinucleotide repeats). Either confirmation of species of origin or presence of known unique cell line markers is considered an adequate test of identity. Expression of the desired product may represent a complementary approach to confirmation of identity.

Microbial cells: For most microbial cells, analysis of growth on selective media is usually adequate to confirm host cell identity at the species level for the host cell bank and the transformed cell bank. For E. coli, where a variety of strains may be used, biological characterisation methods such as phage typing should be considered as supplementary tests of identity. For plasmid banks, identity assessment can be accomplished as described by the ICH document on analysis of the expression construct. Expression of the desired product is also considered adequate to confirm the identity of the microbial expression system.

Tests of purity

A critical aspect of cell development and banking is the assessment that the MCB and WCB are biologically pure, i.e., are free from adventitious microbial agents and adventitious cellular contaminants. The impact of selective agents and antibiotics on the detection of adventious microbial contaminants should be considered when planning and performing these tests.

Metazoan cells: Tests for the presence of bioburden (bacteria and fungi) should be performed on individual containers (1% of the total number but not less than two containers) of the MCB and WCB. In all other aspects, the current methodologies described in either the European Pharmacopoeia (Ph. Eur.), the Japanese Pharmacopoeia (JP) or the U.S. Pharmacopoeia (U.S.P.) for testing microbial limits or microbial sterility may be considered adequate.

Tests for the presence of mycoplasma should be performed on the MCB and WCB. Current procedures considered adequate include both the agar and broth media procedures as well as the indicator cell culture procedure. Current methods for mycoplasma testing are described in Ph. Eur., JP, and «Points to Consider in the Characterization of Cell Lines Used to Produce Biologicals» (FDA, CBER, 1993). Testing cells derived from a single container is generally considered adequate. For non-mammalian animal cell lines, alternative controls and/or assay conditions may be appropriate; manufacturers should consult with the national/regional regulatory authority for appropriate methodology.

If future efforts to harmonise bioburden and mycoplasma assays are fruitful, then the scientifically appropriate harmonised assay should be used.

Virus testing of cell substrates should be designed to detect a wide spectrum of viruses by using appropriate screening tests and relevant specific tests, based on the cultivation history of the cell line, to detect possible contaminating viruses. Applicants should consult the ICH guideline on viral safety. For product classes not covered by the viral safety guideline, the current World Health Organisation (WHO) documents for use of animal cells may be consulted.

The purity of cell substrates can be compromised through contamination by cell lines of the same or different species of origin. The choice of tests to be performed depends upon whether opportunities have existed for cross-contamination by other cell lines. In some cases, it may be necessary to maintain growing cultures of different cell lines in the same laboratory. During procedures in cell banking where open manipulations are performed, care should be taken to ensure that simultaneous open manipulations of other cell lines are avoided to prevent cross-contamination. Whenever another cell line was present in the cell banking room at the same time that open cell banking procedures were being performed (such as cell expansion, pooling, or aliquoting of the chosen cell line), the cell banks should be tested for the presence of cells from (or products derived from) the second cell line. In general, the methods described in «Tests of Identity» to assess cell identity are also considered adequate tests to detect cross-contamination by other cell lines. Additional assurance of lack of cross-contamination can be provided by successful preparation of the intended product from the cell substrate.

Microbial cells: The design and performance of specific tests for adventitious microbial agents and adventitious cellular contaminants in microbial cell banks should take into account the properties of the banked cell, the likely contaminants based upon scientific literature, source, methods and materials used for cultivation, and other organisms present in the banking laboratory. For example, visuel examination of the characteristics of well-isolated colonies is suggested, using several microbiological media, of which some do and some do not support growth of the cell substrate. However, it is not intended that manufacturers necessarily characterise resistant mutants of the cell substrate arising from such studies, or other artifacts of such assays. Rather, the purpose of such assays is to detect existing contaminants.

Cell substrate stability

Another dimension to cell characterisation is appropriateness for intended use in production. There are two concerns for cell substrate stability: Consistent production of the intended product and retention of production capacity during storage under defined conditions.

For the evaluation of stability during cultivation for production, at least two time points should be examined, one using cells which have received a minimal number of subcultivations, and another using cells at or beyond the limit of in vitro cell age for production use described in the marketing application. The limit of in vitro cell age for production use should be based on data derived from production cells expanded under pilot plant scale or commercial scale conditions to the proposed limit of in vitro cell age for production use or beyond. Generally, the production cells are obtained by expansion of cells from the WCB; cells from the MCB could be used with appropriate justification. This demonstration of cell substrate stability is commonly performed once for each product marketing application.

Evaluation of the cell substrate with respect to the consistent production of the intended product of interest should be the primary subject of concern. The type of testing and test article(s) used for such assessments will depend on the nature of the cell substrate, the cultivation methods, and the product. For cell lines containing recombinant DNA expression constructs, consistency of the coding sequence of the expression construct should be verified in cells cultivated to the limit of in vitro cell age for production use or beyond by either nucleic acid testing or product analysis, as described in the relevant ICH guideline. For non-recombinant cell lines in which the coding sequence for the desired product has already been analysed at the MCB or WCB level, invariability of the protein coding sequence during production should be verified in the production cells cultivated to the proposed limit of in vitro cell age for production use or beyond by either nucleic acid testing or analysis of the purified protein product.

Where the product cannot be analysed as described above, other specific traits which may include, for example, morphological characteristics, growth characteristics, biochemical markers, immunological markers, productivity of the desired product, or other relevant genotypic or phenotypic markers may be useful for the assessment of cell substrate stability. In some cases, where direct comparison of the characteristics of the MCB with those of the production cells at or beyond the limit of in vitro cell age is difficult or impossible, one may compare the characteristics of cells at the initial stages of cultivation or production to those of cells at or beyond the limit of in vitro cell age for production use in order to assess cell stability during production. Indices such as, for example, oxygen or glucose consumption rates, ammonia or lactate production rates may be useful for such testing. Increases in the defined limit of in vitro cell age for production use should be supported by data from cells which have been expanded to the proposed new limit of in vitro cell age. For diploid cell lines, data should be presented that establish the finite in vitro lifespan of the cells from the WCB under conditions representative of those employed for manufacturing use.

Evidence for banked cell stability under defined storage conditions will usually be generated during production of clinical trial material from the banked cells. Data from the determination of cell viability when the preserved cells are reconstituted for production of clinical trial supplies will verify that the revived cells have survived the preservation process. Data from the preparation of clinical materials will demonstrate that the revived cells can be used to prepare the desired product. Available data should be clearly documented in the application dossiers, plus a proposal for monitoring of banked cell stability should be provided. The proposed monitoring can be performed at the time that one or more containers of the cryopreserved bank is thawed for production use, when the product or production consistency is monitored in a relevant way, or when one or more containers of the cryopreserved MCB is thawed for preparation of a new WCB (and the new WCB is properly qualified), as appropriate. In the case when production does not take place for a long period of time, viability testing on the cell bank used as a source of the production substrate should be performed at an interval described in the marketing application. If the viability of the cell substrate is not significantly decreased, generally no further testing of the MCB or WCB is considered necessary.

Tests for karyology and tumorigenicity

Utilisation of karyology and tumorigenicity testing for evaluating the safety of a diploid cell line or characterising a new cell line may be useful depending on the cells, the nature of the product and the manufacturing process. Extensive analysis to determine the relative abundance of aneuploid cells has not been found to be useful. Karyology need not be determined for rodent cell lines or new cell lines known to be non-diploid. However, cytogenetic analysis may be an adequate method to assess cell substrate identity or purity as described in «Tests of Identity» and «Tests of Purity». Repetition of tumorigenicity testing for cells with already documented evidence of tumorigenicity is not considered necessary.

For products that are highly purified and that contain no cells, karyology and tumorigenicity testing are generally not considered necessary, provided that appropriate limits for residual host cell DNA are shown to be consistently met by either process validation studies or by lot release testing.

In general, products for which the presence of live cells cannot be excluded or which have little downstream purification (for example, some conventional live virus vaccines) will need such characterisation of the cell substrate. The utility of tumorigenicity testing and chromosomal analysis for new cell substrates for unpurified products should be evaluated on a case-by-case basis. Use of cell lines known to be tumorigenic or to possess abnormal karyology should be evaluated in terms of risk-benefit for each product application when the product contains cells or when not highly purified.

Products that are manufactured in genetically unmodified MRC-5 or WI-38 cells do not need characterisation of these cell substrates by karyology or tumorigenicity since extensive characterisation has already been performed and published for these cell lines. However, for each MRC-5 and WI-38 WCB generated, manufacturers should confirm, once, that the cells grown in the manner to be used in production are diploid and have the expected lifespan.

For new or previously uncharacterised diploid cell substrates, confirmation of diploid karyology should be presented and tumorigenic potential should be established, using cells from the MCB.

Appendix 1: Primary cell substrates .

I. Introduction

The principles contained in this document apply in general to biotechnological/biological products prepared from characterised banked cells. However, a number of biological products, in particular certain viral vaccines, are prepared using primary cells.

Because primary cell cultures are used within the first passage after establishment from the tissue of origin, it is not possible to carry out extensive characterisation of the cells prior to their use as is done for banked cell substrates. In addition, biological products produced using primary cell substrates often do not undergo extensive processing (e.g., purification). Despite these differences, the approach taken to assure the suitability and safety of primary cell substrates for production of biologicals is analogous, in many respects, to that outlined in this document and in other guidelines.

This Annex outlines cell substrate-related information that should be included in marketing applications for biological products prepared using primary cells. This information falls into three general categories: (i) Information concerning the source tissue (or organ) and other animal-derived raw materials used for the establishment of primary cell substrates; (ii) information concerning the preparation of primary cell substrates; and (iii) testing performed on primary cell substrates to ensure the safety of the product.

II. Source tissue and other raw materials

Information should be provided about the animals used as a source of tissue for the preparation of primary cell substrates. Tissue should be derived from healthy animals subjected to veterinary and laboratory monitoring to certify the absence of pathogenic agents. Whenever possible, donor animals should be obtained from closed, specific pathogen-free (when available) colonies or flocks. Animals used as tissue donors should not have been used previously for experimental studies. Animals should be adequately quarantined for an appropriate period of time prior to use for the preparation of cells. In some countries, animals may need to be quarantined in the country where the primary cells are prepared. Manufacturers should consult with national/regional authorities for specific requirements.

Information on materials and components used for the preparation of primary cell substrates should be provided, including the identity and source of all reagents of human or animal origin. A description of testing performed on components of animal origin to certify the absence of detectable contaminants and adventitious agents should be included.

III. Preparation of primary cell substrates

Methods used for isolation of cells from tissue, establishment of primary cell cultures and maintenance of cultures should be described.

IV. Testing of primary cell substrates

Tests performed on primary cell substrates to qualify them for use in production should be described. As noted, the nature of primary cell substrates precludes extensive testing and characterisation prior to use. Testing to demonstrate the absence of adventitious agents in these substrates is therefore conducted concurrently and may include: Observation of production or uninfected control cultures before, during, and beyond the period of production; inoculation of culture fluids from production and uninfected control cultures into various susceptible indicator cell cultures capable of detecting a wide range of relevant viruses, followed by examination for cytopathic changes and testing for the presence of hemadsorbing viruses; and other tests for specific agents (such as relevant retroviruses) as necessary. Additional information concerning specific viral tests may be found in the relevant national/regional/international guidelines.

Appropriate testing regimens and test methods for cells used in the production of specific products will vary depending on the donor species used as a source of tissue, adventitious agents potentially present, the nature of the product, its intended clinical use, aspects of the manufacturing process, and the extent of testing performed on the final product. Applicants should explain and justify the approach taken with respect to their specific product.

Glossary

Cell Bank: A cell bank is a collection of appropriate containers, whose contents are of uniform composition, stored under defined conditions. Each container represents an aliquot of a single pool of cells.

Cell Line: Type of cell population which originates by serial subculture of a primary cell population, which can be banked.

Continuous Cell Line: A cell line having an infinite capacity for growth. Often referred to as «immortal» and previously referred to as «established».

Diploid Cell Line: A cell line having a finite in vitro lifespan in which the chromosomes are paired (euploid) and are structurally identical to those of the species from which they were derived.

Host Cells: See Parental cells.

In Vitro Cell Age: Measure of time between thaw of the MCB vial(s) to harvest of the production vessel measured by elapsed chronological time, by population doubling level of the cells, or by passage level of the cells when subcultivated by a defined procedure for dilution of the culture.

Metazoan: Organism of multicellular animal nature.

MCB (Master Cell Bank): An aliquot of a single pool of cells which generally has been prepared from the selected cell clone under defined conditions, dispensed into multiple containers and stored under defined conditions. The MCB is used to derive all working cell banks. The testing performed on a new MCB (from a previous initial cell clone, MCB or WCB) should be the same as for the MCB unless justified.

Parental Cells: Cell to be manipulated to give rise to a cell substrate or an intermediate cell line. For microbial expression systems, it is typical to also describe the parental cells as the host cell. For hybridomas, it is typical to also describe the parental cells as the cells to be fused.

WCB (Working Cell Bank): The Working Cell Bank is prepared from aliquots of a homogeneous suspension of cells obtained from culturing the MCB under defined culture conditions.

List of Participants

ADAMSON R., Genetics Institute, 1 Burtt Road, Andover, MA 01810, USA.

ARMAND J., Pasteur Mérieux Connaught, 1541, avenue Marcel Mérieux, F-69280 Marcy L'Etoile.

BAECKMANN A., Bundesamt für Arzneimittel und Medizine Produckt, Seestrasse 10, D-13353 Berlin.

BARON H., Rhone Poulenc Rorer, Drug Safety Division, 500 Arcola Road, Box 1200, Collegeville, PA 19426-0107, USA.

BERTHOLD W., Hoffman-Laroche Inc., 340 Kingsland Street, Nutley, NJ 07110-1199, USA.

BEUVERY E.C., RIVM, Antonie van Leeuwenhoeklaam 9, Postbus 1, NL-3720 BA Bilthoven.

BRADLEY R., Central Veterinary Laboratory, BSE Co-Ordinator, Woodhand Lane, New Haw, Addelstone, Surrey KT15 3NB, U.K.

BROWN F., Plum Island Animal Disease Center, P.O. Box 848, Greenport, NY 11944, USA.

BROWN P., Laboratory of Central Nervous System Studies, Building 36, Room 5B21, Bethesda, MA 20892, USA.

BUILDER S., Genentech, Inc., 460, Point San Bruno Blvd, South San Francisco, CA 94080, USA.

CHIU Y.-Y., US Food and Drug Administration, Rm/4B32, 5600 Fishers Lane, Rockville, MD 20857, USA.

DE LUCA G.P., Ares Services S.A., Chemin des Mines 15 bis, CH-1202 Geneva.

DOBBELAER R., Ministry of Public Health, Institute for Hygiene & Epidemiology, rue J. Wystman 14, B-1050 Bruxelles.

DODET B., Fondation Marcel Mérieux, 17, rue Bourgelat, B.P. 2021, F-69227 Lyon Cedex 02.

DORNER F., Immuno A.G., Uferstrasse 15, A-2304 Orth/Donau.

EIBL J., Immuno AG, Research and Development, Industriestrasse 67, A-1220 Vienne.

ELSKEN L.A., U.S. Department of Agriculture, National Veterinary Services Laboratories, 1800 Dayton Road, P.O. Box 844, Ames, Iowa 50010, USA.

FARMER G., Ares Serono, Corporate Director Regulatory Affairs, 15, chemin des Mines, CH-1202 Geneva.

FELDMAN F., Pre-clinical R&D Biotech Evaluation, Centeon, 1020, First avenue, King of Prussia, PA 19406-1310, USA.

FLETCHER M., Pasteur Mérieux Connaught, 3, avenue Pasteur, F-92430 Marnes la Coquette.

FURMINGER I.G., Evans Medical Ltd, Evans House, Regent Park, Kingstone Road, Leatherhead, Surrey KT22 7PQ, U.K.

GARFINKLE S., Merck & Co., Inc., Sumneytown Pike, WP36M-6, P.O. Box 4, West Point, PA 19486, USA.

GARNICK R., Genentech Inc., 1, DNA Way, South San Francisco, CA 94080-4990, USA.

GRACHEV V., Institute of Poliomyelitis and Viral Encephalitides, Russian Academy of Medical Science, 142782 Moscou Region, Russia.

GRIFFITHS E., Biologicals Unit, WHO, CH-1211 Geneva 27.

GUST I., CSL Ltd, 45 Poplar Road, Parkville, Victoria 3052, Australia.

HARDEGREE C., Food and Drug Administration, Center for Biologics Evaluation and Research, 1401 Rockville Pike, Suite 200N, Rockville, MD 20852-1448, USA.

HAY R., American Type Culture Collection, 10801 University Blvd, Manassas, VA 20110, USA.

HAYAKAWA T., National Institute of Health Science, 1-18-1 Kamiyoga, Setagayaku, Tokyo 158, Japan.

HERBERT A., Connaught Laboratories Limited, 1755 Steeles Avenue West, North York, Ontario M2R 3T4, Canada.

HESSEL L., Pasteur Mérieux Sérums et Vaccins, 58, avenue Leclerc, F-69007 Lyon.

HIMES V., Wyeth-Lederle Vaccines & Pediatrics, Wasp and Biddle Streets, P.O. Box 304, Marietta, PA 17547, USA.

HORAUD F., Institut Pasteur, 25, Rue du Docteur Roux, F-75724 Paris Cedex 15.

HUGHES B.P., Glaxo Wellcome Research and Development, Langley Court, South Eden Park Road, Beckenham, Kent BR3 3BS, U.K.

KAWANISCHI, National Institute of Health Science, 1-18-1 Kamiyoga, Setagayaku, Tokyo 158, Japan.

KRAUSE P.R., CBER/FDA, Office of Vaccine Research and Review, HFM-457, Bldg N29A, Room IC18, 1401 Rockville Pike, Rockville, MD 02852, USA.

LACOSTE F., Merial, 254, rue marcel Mérieux, F-69342 Lyon Cedex 07.

LAMY D., Transgène, 11, rue de Molsheim, F-67000 Strasbourg.

LARZUL D., Institut Pasteur, Dir. Applications de la Recherche, 28, rue du Docteur Roux, F-75724 Paris Cedex 15.

LECHENET J., Merial, 254, rue Marcel Mérieux, F-69342 Lyon Cedex 07.

LEWIS A., Office of Vaccines Research and Review/CBER/FDA HFM-400, Bldg N29B, Room 1NN24, 1401 Rockville Pike, Rockville, MD 02852-1448, USA.

LUBINIECKI T., SmithKline Beecham Pharmaceuticals, 709 Swedeland Road, P.O. Box 1539 – UE-0428, King of Prussia, PA 19406-0939, USA.

LUPKER J.H., Sanofi Recherche, Labège Innopole, Voie N° 1, BP 137, F-31676 Labège Cedex.

MAHY B., Center for Diseases Control, National Immune Programme, 1600 Clinton Road, Atlanta, GA 30353, USA.

MARTIN J.-F., Biocine, 36, quai Fulchiron, F-68005 Lyon.

MATTIS J., Centocor, 2000 Great Valley Parkway, Malvern, PA 19355-1307, USA.

McGARRITY G.J., Genetic Therapy, 938 Clopper Road, Gaithersburg, MD 20878, USA.

MERIEUX C., Fondation Marcel Mérieux, 17, rue Bourgelat, B.P. 2021, F-69227 Lyon Cedex 02.

MEULIEN P., Research & Development, Pasteur Mérieux Connaught, 1755 Steeles Ave. West, North York, Ont. M2R 3T4, Canada.

MINOR P., NIBSC, Blanche Lane, South Mimms, Potters Bar, Herts EN6 3QG, U.K.

MONTAGNON B., Pasteur Mérieux Connaught, 1541, avenue Marcel Mérieux, F-69280 Marcy L'Etoile.

MORANDI M., Chiron Biocine, 1, via Florentina, I-53100 Sienna.

MORIN O., International Federation of Pharmaceutical Manufacturers Assocation, 30, rue de St-Jean, CH-1211 Geneva 18.

MORSE S., The Rockefeller University, 1230 York Avenue, Box 173, New York, NY 10021-6399, USA.

MURANO G., FDA/CBER/OTRR/DHP/29A/2D16, 1401 Rockville Pike (HFM 535), Rockville, MD 20852-1448, USA.

NICHOLS W., Merck & Company, Merck Research Labs WP 45-204, West Point, PA 19486, USA.

OLIVA O.P., Biomanguinhos, Oswaldo Cruz Foundation, avenue Brasil, 4365, CEP 21.045-900, Rio De Janeiro, RJ, Brazil.

ONIONS D.E., Department of Veterinary Pathology, University of Glasgow Veterinary School, Beardsen Road, Glasgow G61 1QH, U.K.

PEETERMANS J., SmithKline Beecham Biologicals SA, 89, rue de l'Institut, B-1330 Rixensart.

PETRICCIANI J., P.O. Box 1925, Palm Springs, CA 92263-1925, USA.

PLOTKIN S., Pasteur Mérieux Sérums et Vaccins, 3, avenue Pasteur, F-92430 Marnes-La-Coquette.

RIDGWAY A.G., Health and Welfare, Biotechnology Section, Blood Products Division, Tunney's Pasture 0700E, Building 7, Ottawa, Ontario K1A 0L2, Canada.

ROBERTSON J., NIBSC, Blanche Lane, South Mimms, Potters Bar, Herts EN6 3QG, U.K.

ROUMIANTZEFF M., 1, rue Dangon, F-69004 Lyon.

SAKAMOTO H., Asahi Chemical Industry Co. Ltd, 4-5-13, Shibaura, Minato-Ku, Tokyo 108, Japan.

SCHILD G., NIBSC, Blanche Lane, South Mimms, Potters Bar, Herts EN6 3QG, U.K.

SEAMON K.B., Immunex Corporation, 51, University Street, Seattle, WA 98101-2936, USA.

SINOGAS C., Instituto Gulbenkian De Cienca, rua da Quinta Grande, 6 – Apartado 14, P-2781 Oeiras Codex.

SPIESER J.M., Secrétariat de la Pharmacopée Européenne, 224-226 avenue de Colmar, F-67100 Strasbourg.

STRUCK M.M., Swiss Serum & Vaccine Institute, Rehhagstrasse 79, CH-3018 Berne.

VALENTIN C., Pasteur Mérieux Connaught, 1541, avenue Marcel Mérieux, F-69280 Marcy L'Etoile.

VYAS G.N., University of California, San Francisco, Department of Laboratory Medicine, 513 Parnassus avenue, S-535, San Francisco, CA 94143, USA.

WEISS R.A., The Institute of Cancer Research, Chester Beatty Laboratories, 237 Fuhlam Road, London SW3 6JB, U.K.

WILL R.G., Western General Hospital, Department of Clinical Neurosciences, Crewe Road, Edinburgh, EH4 2XU, U.K.

WOLFF R., FDA, 1401 Rockville Pike, Rockville, MD 20852-1448, USA.

WUDL L., Amgen, 3200 Walnut Street, Thousand Oaks, CA 91320, USA.

ZHOU H., National Institute for the Control of Pharmaceutical and Biological Products, Temple of Heaven, Beijing, République of China.

ZORZI-MORRE P., Sanofi Recherche/Centre de Labège, Développement en Biotechnologie (Bureau N-218), B.P. 137, F-31676 Labège Cedex.

Index of Authors

Symposia Series in Immunobiological Standardization

Vol. 1: Talloires 1965	Rabies
Vol. 2: Munich 1965	Neurovirulence of viral vaccines
Vol. 3: Stockholm 1965	Biotechnical developments in bacterial vaccine production
Vol. 4: Royaumont 1965	Immunological methods of biological standardization
Vol. 5: London 1966	Laboratory animals
Vol. 6: Utrecht 1966	Adjuvants of immunity
Vol. 7: Marburg/L. 1967	Assay of combined antigens
Vol. 8: Lyon 1967	Foot-and-Mouth Disease: Variants and immunity
Vol. 9: Paris 1967	Pseudotuberculosis
Vol. 10: London 1967	Biological assay methods of vaccine and sera
22nd Symp., London 1968	Standardization of immunofluorescence. Ed. Blackwell
Vol. 11: London 1968	Rubella vaccines
Vol. 12: Tunis 1968	Brucellosis. Standardization and control of vaccines and reagents
Vol. 13: Utrecht 1969	Pertussis vaccine
Vol. 14: London 1969	Standardization of interferon and interferon inducers
Vol. 15: Berne 1969	Enterobacterial vaccines
Vol. 16: Versailles 1970	Anti-lymphocytic serum
29th Symp., Brighton 1969	Hazards of handling simians. Ed. Lab. Animals Ltd. (4)
Vol. 17: Frankfurt 1970	BCG vaccine
Vol. 18: Copenhagen 1972	HL-A reagents
Vol. 19: Utrecht 1972	Smallpox vaccines
Vol. 20: London 1972	Influenza
Vol. 21: Lyon 1972	Rabies II
Vol. 22: Monaco 1973	The use of vaccines

Developments in Biological Standardization

Vol. 23: Madrid 1973	Sterilization and sterility testing of biological substances
Vol. 24: San Francisco 1973	Preservatives in biological products
Vol. 25: Lyon 1973	Requirements for poultry virus vaccines
Vol. 26: Budapest 1973	Selected veterinary vaccines
Vol. 27: Budapest 1973	Purity of human plasma proteins
Vol. 28: London 1974	Immunity to infections of the respiratory system in man and animals
46th Symp., London 1974	Breeding of simians and their uses in developmental biology. Ed. Lab. Animals Ltd. (6)
Vol. 29: Geneva 1974 (WHO/IABS)	Standardization and control of allergens administered to man
Vol. 30: Milan 1974	Viral hepatitis
Vol. 31: Rabat 1975	Brucellosis (II)
Vol. 32: Paris 1975 (OIE/IABS)	Colstridial sera and vaccines for veterinary use
Vol. 33: Douglas 1975	Vaccination by the non-parenteral route
Vol. 34: Budapest 1976	Pyrogenicity, innocuity and toxicity test systems for biological products
Vol. 35: Lyon 1976	Foot-and-mouth disease vaccines

Vol. 36: Lyon Washington D.C. 1976	Freeze-drying of biological products
Vol. 37: Geneva 1976	Standardization of cell substrates for the production of virus vaccines
Vol. 38: London 1977	Biological preparations in the treatment of cancer
Vol. 39: Geneva 1977	Influenza immunization (II)
Vol. 40: Marburg 1977	Standardization of rabies vaccines (III) (WHO/IABS)
Vol. 41: Guadeloupe 1978	Vaccinations in the developing countries
Vol. 42: Paris 1978	Second General Meeting of European Society of Animal Cell Technology
Vol. 43: Brussels 1978	Immunization: Benefit versus risk factors
Vol. 44: Villars 1979	Test methods for the quality control of plasma proteins
Vol. 45: San Antonio 1979	The Standardization of animals to improve biomedical research, production and control
Vol. 46: Oxford 1979	Third General Meeting of European Society of Animal Cell Technology
Vol. 47: Bilthoven 1980	Reassessment of inactivated poliomyelitis vaccine
Vol. 48: Geneva 1980	Standardization of albumin, plasma substitutes and plasmapheresis
Vol. 49: Leetown 1981	Fish biologics: Serodiagnostics and vaccines
Vol. 50: Heidelberg 1981	Joint ESACT-IABS Meeting on the use of heteroploid and other cell substrates for the production of biologicals
Vol. 51: Lyon 1981	Immunization of adult birds with inactivated oil adjuvant vaccines
Vol. 52: Lyon 1981	Herpes virus of man and animal: Standardization of immunological procedures
Vol. 53: Dublin 1982	Enteric infections in man and animals: Standardization of immunological procedures
Vol. 54: Athens 1982	Viral hepatitis: Standardization in immunoprophylaxis of infections by hepatitis viruses
Vol. 55: Copenhagen 1982	Fifth General Meeting of European Society of Animal Cell Technology
Vol. 56: Algeria 1983	IIIrd International Symposium on Brucellosis
Vol. 57: Paris 1983	Monoclonal antibodies: Standardization of their characterization and use
Vol. 58: Budapest 1983	BCG vaccines and tuberculins
Vol. 59: Geneva 1983	Standardization and control of biologicals produced by recombinant DNA technology
Vol. 60: Gardone Riviera 1983	Production and exploitation of existing and new animal cell substrates
Vol. 61: Geneva 1984	Proceedings of the Fourth International Symposium on pertussis
Vol. 62: Stockholm 1985	Diagnostics and vaccines for parasitic diseases
Vol. 63: San Francisco 1985	Use and standardization of chemically defined antigens
Vol. 64: London 1985	Reduction of animal usage in the development and control of biological products
Vol. 65: Amsterdam 1985	Use and standardization of combined vaccines
Vol. 66: Baden 1985	Advances in animal cell technology: cell engineering, evaluation and exploitation
Vol. 67: Melbourne 1986	Standardization in blood fractionation including coagulation factors
Vol. 68: Geneva 1986	Cells – Products – Safety
Vol. 69: London 1987	Cytokines: Laboratory and clinical evaluation
Vol. 70: Arlington, 1988	Continuous cell lines as substrates for biologicals
Vol. 71: Utrecht, 1989	Monoclonal antibodies for therapy, prevention and *in vivo* diagnosis of human disease
Vol. 72: Annecy, 1989	Progress in animal retroviruses
Vol. 73: Shizuoka, 1990	Pertussis: Evaluation and research on acellular pertussis vaccines
Vol. 74: Bethesda, MD, 1990	Biological product freeze-drying and formulation
Vol. 75: London, 1990	Virological aspects of the safety of biological products
Vol. 76: Bethesda, MD, 1991	Continuous cell lines – Current issues
Vol. 77: Annecy, 1991	Standardization of the immunopharmacology of natural and synthetic immunomodulators
Vol. 78: Bethesda, MD, 1991	Poliovirus attenuation: Molecular mechanisms and practical aspects
Vol. 79: Ploufragan, 1992	The first steps towards an international harmonization of veterinary biologicals: 1993 and free circulation of vaccines within the E.E.C.
Vol. 80: Heidelberg, 1992	Transmissible spongiform encephalopathies-Impact on animal and human health.
Vol. 81: Cannes, 1992	Virological safety aspects of plasma derivatives.
Vol. 82: Albany, NY, 1993	Recombinant vectors in vaccine development.
Vol. 83: Annecy, 1993	Genetic stability and recombinant product consistency.

Vol. 84: Langen, 1993	Non-target effects of live vaccines.
Vol. 85: Santa Fee, NM, 1994	Genetics of streptococci, enterococci and lactococci.
Vol. 86: Langen, 1994	Replacement, reduction and refinement of animal experiments in the development and control of biological products.
Vol. 87: Geneva, 1995	New approaches to stabilisation of vaccines potency.
Vol. 88: Bethesda, MD, 1995	Viral safety and evaluation of viral clearance from biopharmaceutical products.
Vol. 89: Rome, 1995	Pertussis vaccine trials.
Vol. 90: Oslo, 1996	Fish vaccinology.
Vol. 91: San Francisco CA, 1996	Development of specifications for biotechnology pharmaceutical products.
Vol. 92: Bergen, 1996	Modulation of the immune response to vaccine antigens.

INTERNATIONAL ASSOCIATION OF BIOLOGICAL STANDARDIZATION

WHAT IS THE I.A.B.S. ?

The International Association of Biological Standardization (IABS) was founded in Lyons in 1955 by a group of independent experts who identified an urgent need for an improvement in the quality and comparability of the data being exchanged between scientists working in the research, development, production, standardization and regulation of biological products. To this end, IABS aims to bring together state controllers, manufacturers and research workers from academic institutions and public health organisations.

The IABS is a branch of the International Union of Microbiological Societies and has about 500 members in 50 countries. Every four years, its members elect the Council by postal ballot and its representatives, in their turn, elect the officers.

The duty of the Council is to organise meetings on subjects of contemporary interest in biological products being used in human and veterinary medicine and to promote the editorial activity of the IABS by publishing: i) the proceedings of the international meetings in the symposium series *Developments in Biological Standardization* published by S. Karger, ii) *Biologicals,* a major international journal published quarterly by Academic Press (London), and iii) the *Newsletter* which reports on current IABS activities.

The Council also supports the objectives of the major international public health programmes and organisations, and particularly the World Health Organization.

In 1994, the Council decided to establish an IABS Task Force on Vaccines to provide a focus for our activities in the field of vaccination and immunisation. The role of the Task Force is to foster communication between vaccine manufacturers, scientific research workers and regulatory authorities concerning the quality control, standardization and efficacy of vaccines and vaccine combinations and the exchange of information on the development of new improved vaccines. A major objective of the Task Force will be the organisation of international meetings and the promotion of high quality publications on vaccines in our journal *Biologicals* and in the *Developments in Biological Standardization.*

The Task Force has a membership of 30 leading international experts on vaccination and immunisation working together to identify the major topics in the field of vaccines that need international debate and cooperation.

INTERNATIONAL ASSOCIATION OF BIOLOGICAL STANDARDIZATION

WHY NOT BECOME A MEMBER OF THE IABS?

As a member of the IABS, it is possible to contribute directly to the activity of our Association and to support the research, development, standardization and regulation of biological products. By your membership you would:

- receive our quarterly *Newsletter* which gives information on the activities of the IABS and highlights international news of current interest;
- subscribe to our journal *Biologicals*, published quarterly by Academic Press (London), at a reduced rate;
- take advantage of the discount on booksellers' prices for the volumes in the symposium series *Developments in Biological Standardization*, published by Karger;
- participate fully in our international meetings, with a reduction in the registration fees.

If you are not already a member of our Association and wish to join us, please complete this form and send it to our permanent office:

BIOSTANDARDS
P.O.B. 456
78, av. de la Roseraie
CH-1211 Geneva 4
Switzerland
Fax: (41) 22 702 93 55

or hand it to the Secretariat, if attending a Meeting.

- -

Surname...First name(s)..

Scientific titles...Professional position..

Complete mailing address..

....

Field of expertise...

Tel... Fax...

Annual fee: Sfr. 180.– which includes membership, subscription to *Biologicals* and the quarterly *Newsletter* of the IABS;
or Sfr. 90.-, without the subscription to *Biologicals*.

Method of payment:

❍ I have given instructions to my bank to make a credit transfer to the Swiss Bank Corporation, P.O.B. CH-1211 Geneva 11, Switzerland, a/c Biostandards n° 131,652.0.
❍ I enclose my cheque for Sfr................................ made payable to Biostandards
❍ I authorise you to charge Sfr................... to my card: VISA / MASTERCARD / AM. EXPRESS

N°.. Expiry date........................

Signature..

IMPRIMERIE MEDECINE m+h HYGIENE
juillet - 1998